An Introduction to the Philosophy of Mathematics

This introduction to the philosophy of mathematics focuses on contemporary debates in an important and central area of philosophy. The reader is taken on a fascinating and entertaining journey through some intriguing mathematical and philosophical territory. Topics include the realism/anti-realism debate in mathematics, mathematical explanation, the limits of mathematics, the significance of mathematical notation, inconsistent mathematics, and the applications of mathematics. Each chapter has a number of discussion questions and recommended further reading from both the contemporary literature and older sources. Very little mathematical background is assumed, and all of the mathematics encountered is clearly introduced and explained using a wide variety of examples. The book is suitable for an undergraduate course in philosophy of mathematics and, more widely, for anyone interested in philosophy and mathematics.

MARK COLYVAN is Professor of Philosophy and Director of the Sydney Centre for the Foundations of Science at the University of Sydney. He is the co-author (with Lev Ginzburg) of *Ecological Orbits: How Planets Move and Populations Grow* (2004) and author of *The Indispensability of Mathematics* (2001).

An Introduction to the Philosophy of Mathematics

MARK COLYVAN

University of Sydney

CAMBRIDGE
UNIVERSITY PRESS

CAMBRIDGE
UNIVERSITY PRESS

University Printing House, Cambridge CB2 8BS, United Kingdom

Published in the United States of America by Cambridge University Press, New York

Cambridge University Press is part of the University of Cambridge.

It furthers the University's mission by disseminating knowledge in the pursuit of education, learning and research at the highest international levels of excellence.

www.cambridge.org
Information on this title: www.cambridge.org/9780521826020

© Mark Colyvan 2012

First published 2012
Reprinted 2013

A catalogue record for this publication is available from the British Library

Library of Congress Cataloguing in Publication data
Colyvan, Mark.
  An introduction to the philosophy of mathematics / Mark Colyvan.
     p. cm
  Includes bibliographical references and index.
  ISBN 978-0-521-82602-0 (hardback) – ISBN 978-0-521-53341-6 (paperback)
  1. Mathematics–Philosophy.   I. Title.
  QA8.4.C654 2012
  510.1–dc23   2012007499

ISBN 978-0-521-82602-0 Hardback
ISBN 978-0-521-53341-6 Paperback

Contents

Acknowledgments

I'd like to start by thanking those who taught me most of what I know about the philosophy of mathematics. I have benefitted enormously from both the written work of and conversations with: John Bigelow, Jim Brown, Hartry Field, Drew Khlentzos, Penelope Maddy, Mike Resnik, Stewart Shapiro, and Mark Steiner. Others with whom I've had many interesting conversations on at least some of the topics covered in this book include: Jody Azzouni, Alan Baker, J. C. Beall, Otávio Bueno, Colin Cheyne, Alan Hájek, Chris Hitchcock, Mary Leng, Ed Mares, Bob Meyer, Chris Mortensen, Daniel Nolan, Graham Priest, Greg Restall, Jack Smart, and Ed Zalta.

All of the people just mentioned have been very influential on my thinking on the topics covered in this book. Their ideas appear scattered throughout this book as part of my background knowledge, as points of departure for my own views, and as prominent positions in the intellectual landscape. Without their contributions, this book would simply not have been possible.

I'd also like to thank my students. I have taught undergraduate and graduate courses based on the material in this book at a number of universities in Australia and the USA. The students in these courses have forced me to be clearer and more rigorous in my presentation. This, in turn, has greatly improved the way I teach the material and very often has led to refinements in my thinking about the issues in question. In many ways this book is my attempt to meet the exacting standards of my students.

Some of the material for this book is drawn from previously published material. Chapter 3 uses material from my article: M. Colyvan, 1998, 'Indispensability Arguments in the Philosophy of Mathematics', in E. N. Zalta (ed.), *The Stanford Encyclopedia of Philosophy*, online edition, http://plato.stanford.edu/entries/mathphil-indis/. I thank the editors for permission to reproduce that material here. Chapter 4 makes use of material from M. Colyvan, 2010, 'There Is No Easy Road to Nominalism', *Mind*, 119(474): 285–306, and M. Colyvan, 2011, 'Fictionalism in the Philosophy of Mathematics', in E. J. Craig (ed.), *Routledge Encyclopedia of Philosophy*, online edition, www.rep.routledge.com/article/Y093. I thank the editors of *Mind* and the *Routledge Encyclopedia of Philosophy*, for permission to reproduce that material here, and I note that the copyright on this material remains with Oxford University Press and Routledge, respectively. Finally, Chapter 7 draws on material from M. Colyvan, 2008, 'Who's Afraid of Inconsistent Mathematics?', *Protosociology*, 25: 24–35. I am grateful to the editor of *Protosociology* for permission to reproduce the material in question here; the copyright remains with that journal.

I am also indebted to a reader for Cambridge University Press who read an earlier draft of this book and made many extremely helpful comments and suggestions. The end result is greatly improved thanks to this reader's efforts.

I owe special thanks to Bill Newell, who is largely responsible for my ongoing interest in mathematics. Apart from anything else, Bill taught me to appreciate the extraordinary beauty of mathematics. I hope that some of what I learned from him shines through in the pages that follow.

1 Mathematics and its philosophy

Mathematics is the queen of the sciences and arithmetic is the queen of mathematics. She often condescends to render service to astronomy and other natural sciences, but in all relations, she is entitled to first rank.

Carl Friedrich Gauss (1777–1855)[1]

Mathematics occupies a unique and privileged position in human inquiry. It is the most rigorous and certain of all of the sciences, and it plays a key role in most, if not all, scientific work. It is for such reasons that the great German mathematician Carl Friedrich Gauss (1777–1855) pronounced mathematics to be the queen of the sciences. But the subject matter of mathematics is unlike that of any of the other branches of science. Mathematics seems to be the study of mathematical entities – such as numbers, sets, and functions – and the structural relationships between them. Mathematical entities, if there are such things, are very peculiar. They are abstract: they do not have spatiotemporal location and do not have causal powers. Moreover, the methodology of mathematics is apparently unlike the methodology of other sciences. Mathematics seems to proceed via a-priori means using deductive proof, as opposed to the a-posteriori methods of experimentation and induction found in the rest of science. And, on the face of it at least, mathematics is not revisable in the way that the rest of our science is. Once a mathematical theorem is proven, it stands forever. Mathematics may well be the queen of the sciences, but she would seem to be an eccentric and obstinate queen.

The philosophy of mathematics is the branch of philosophy charged with trying to understand this queen. We investigate the limits of mathematics, the subject matter of mathematics, the relationship between mathematics

---

[1] Sartorius von Waltershausen, *Gauss zum Gadächtniss*, 1856, p. 79. Quoted in Robert Edouard Moritz, *Memorabilia Mathematica*, 1914, p. 271.

and the rest of science, the logic of mathematical proofs, and the significance of the language of mathematics to mathematical practice. These are all important topics, and we address each of them in this book. They are significant for both philosophy and for mathematics. For example, understanding one of the paradigmatic cases of secure, a-priori knowledge is crucial to the branch of philosophy concerned with knowledge and its acquisition: epistemology. The importance of philosophy of mathematics to mathematics is also clear. Apart from anything else, philosophy sheds light on what mathematics is about. No self-respecting branch of science should be in the position of not knowing what its primary object of study is. More importantly, it may well be that the very methodology of mathematics hangs on the answers to some of the philosophical questions that impose themselves upon us. A brief look at the history of the relationship between mathematics and philosophy of mathematics will help illustrate the importance of philosophy of mathematics for both philosophy and mathematics.

1.1 Skipping through the big isms

The first half of the twentieth century was a golden age for philosophy of mathematics. It started with a philosopher, Bertrand Russell (1872–1970), proving that the foundational mathematical theory, set theory, was inconsistent. This led to a crisis in the foundations of mathematics and an intense period of debate. The debate and subsequent development of new set theories involved major philosophers of the time, such as Frank P. Ramsey (1903–30), Ludwig Wittgenstein (1889–1951), Gottlob Frege (1848–1925), Edmund Husserl (1859–1938), Charles Sanders Peirce (1839–1914), and of course Russell and his collaborator Alfred North Whitehead (1861–1947). Leading figures in mathematics were also involved. These included Hermann Weyl (1885–1955), Henri Poincaré (1854–1912), Kurt Gödel (1906–78), David Hilbert (1862–1943), L. E. J. Brouwer (1881–1966), Ernst Zermelo (1871–1953), and Alfred Tarski (1901–83).[2] The participants in these debates are major figures and household names (in my household, at least). There

---

[2] The distinction between philosophers and mathematicians here is somewhat arbitrary; many of these people should rightly be thought of as both philosophers and mathematicians. And, of course, there were many other major figures involved in these debates –

is no doubt about it, these must have been heady times – times when philosophy of mathematics really mattered, and everybody knew it.

Sadly, the excitement of these times didn't last. The debates over the foundations of mathematics bogged down. After a very productive 30 or 40 years, very little progress was made thereafter, and, by and large, both philosophers and mathematicians became tired of the philosophy of mathematics. At least, they became tired of the major movements of the first half of the twentieth century – 'the big isms' we'll get to shortly – and purely foundational issues in mathematics. Philosophy of mathematics kept going, of course – philosophy always does – but it had lost its urgency and, to some extent, its *raison d'être*.

It is very easy, as a student of philosophy of mathematics, to spend one's time looking back to the debates and developments of the first half of the twentieth century. But the philosophy of mathematics has moved on, and it is once again relevant and engaged with mathematical practice. The aim of this book is to get beyond the first half of the twentieth century and explore the issues capturing the attention of contemporary philosophers of mathematics. I will thus relegate most of the historical material to this short section, where we look at three of 'the big isms', and to the following chapter.[3] In Chapter 2 we consider some of the important mathematical results about the limits of mathematics. Although most of the results are from the first half of the twentieth century, they still loom large in contemporary philosophy of mathematics and thus deserve a more extensive treatment. My apologies to anyone who is disappointed by the relatively superficial treatment of the early twentieth-century philosophy of mathematics. While very good discussions of these topics abound, entry-level accounts of contemporary philosophy of mathematics are rare.

Below I give the briefest outline of three of the major movements in the philosophy of mathematics from the early twentieth century. Each of these has its charms; they each take one particular aspect of mathematical methodology as central to understanding mathematics. I should add that the three positions outlined below are historically very important, but they

---

too many to list here. The interested student is encouraged to read about the relevant history; it is a fascinating story, involving many noteworthy characters.

[3] The fourth 'ism', Platonism, is still very prominent in the contemporary literature so earns a chapter to itself: Chapter 3.

are not merely of historical interest – there are modern defenders of versions of each of them. It's just that the discussions of these positions no longer take centre stage.

1.1.1 Formalism

This view takes mathematical notation and its manipulation to be the core business of mathematics. In its purest form, formalism is the view that mathematics is nothing more than the manipulation of meaningless symbols. So-called *game formalism* is the view that mathematics is much like chess. The pieces of a chess set do not represent anything; they are just meaningless pieces of wood, metal, or whatever, defined by the rules that govern the legal moves that they can participate in. According to game formalism, mathematics is like this. The mathematical symbols are nothing more than pieces in a game and can be manipulated according to the rules. So, for example, elementary calculus may tell us that $d(ax^2 + bx + c)/dx = 2ax + b$. This is taken by formalism to mean that the right-hand side of the equation can be reached by a series of legal mathematical 'moves' from the left-hand side. As a result of this, in future mathematical 'games' one is licensed to replace the symbols '$d(ax^2 + bx + c)/dx$' with the symbols '$2ax + b$'. That too becomes a legal move in the game of mathematics. There are more sophisticated versions of formalism, but that's the basic idea. There is a question about whether the 'pieces' of the game are the actual mathematical symbol tokens, or whether it is the symbol types. That is, is this instance of 'π' different from, or the same as, this one: 'π'? They are two different tokens of the same type. Formalists need to decide where they stand on this and other such issues. Different answers give rise to different versions of formalism.

Formalism faces a number of difficulties, including accounting for the usefulness of mathematics in applications. But for now we just want to get a sense of what formalism is and why it was, at one time, a serious contender as a philosophy of mathematics. For a start, and as I've already mentioned, formalism takes notation seriously.[4] Indeed, it takes mathematics as being primarily about the notation. In so doing, it avoids problems associated with other accounts of mathematics, whereby the notation is

---

[4] See Chapter 8 for more on the importance of notation to mathematics.

taken to be standing for mathematical objects.[5] Formalism also places great importance on stating what the legal manipulations of the symbols are and which symbols are legitimate. This approach sits very well with a great deal of mathematics, especially axiomatic theories such as set theory and group theory.[6] The axioms of these theories function as the specification of both the legal manipulations in question and the objects of manipulation. And the formalist's suggestion that there is nothing more to these theories is not altogether mad. For example, in set theory the membership relation $\in$ really does seem to be a primitive notion, defined implicitly by the theory in which it resides. Just as the question of what a bishop in chess is can be answered in full by explaining the rules of chess and the role a bishop plays in the game. There is nothing more to say in either case, or so goes the formalist line of thought. As we shall see in the next chapter, it is generally thought that the most sophisticated version of a theory along these lines was put to rest by Gödel's Incompleteness Theorems. In any case, formalism has few supporters these days. But the other big isms are in better shape.[7]

1.1.2 Logicism

This view of mathematics takes the a-priori methodology of mathematics as central. According to logicism, mathematics is logic. That's the slogan, at least; spelling out what this slogan amounts to is more difficult, but the basic idea is that mathematical truths can, in some sense, be reduced to truths about logic. The position is epistemologically motivated: logical knowledge is thought to be more basic and less mysterious than mathematical knowledge. Given the German mathematician Richard Dedekind's (1831–1916) reduction of real numbers to sequences of rational numbers[8] and other known reductions in mathematics, it was tempting to see basic arithmetic as the foundation of mathematics. Moreover, if arithmetic were

---

[5] We will encounter the problems with such realist accounts of mathematics in due course.

[6] See p. 88 for the axioms of group theory.

[7] See Curry (1951) for a classic defence of formalism and Weir (2010) for an interesting modern attempt to resuscitate the position.

[8] Dedekind's idea was to identify real numbers with the limits of sequences of rational numbers – so-called *Dedekind cuts*.

to turn out to be derivable from logic, then we'd have a compelling account of the nature of mathematics. Logicism was first proposed and developed in detail by Gottlob Frege. Unfortunately Frege system was inconsistent. He included the now-infamous Basic Law V as one of his logical axioms.[9] This rather innocuous-looking axiom about the extensions of predicates was shown by Bertrand Russell to lead to a contradiction. But many thought that Frege was onto something. Indeed, Russell was one of them. He, in collaboration with Whitehead, pushed the logicist programme forward, but the further this programme was developed, the less the basic machinery looked as if it deserved to be called 'logic'.

The allure of logicism and the considerable achievements of Frege live on, though. The contemporary descendant of this programme is *neo-logicism*. The neo-logicist programme takes as its point of departure the fact that Frege did not really need anything so strong as his problematic Basic Law V in order to get most of what he wanted. Basic Law V can be replaced with Hume's principle: the number of Fs is equal to the number of Gs iff the Fs and the Gs can be placed in one–one correspondence. Strictly speaking, this principle is not a law of logic, but it's very, very close. With Hume's principle in hand and helping oneself to second-order logic,[10] the core of Frege's project can be carried out.[11]

1.1.3 Intuitionism

This view of mathematics takes proof in mathematics very seriously. Indeed, according to intuitionism, proof and constructions are all there is. (Intuitionism is sometimes called *constructivism* for this reason.) Accordingly, mathematics is not taken to be about some pre-existing realm of mathematical objects. Intuitionism has it that mathematical objects need to be constructed before one can sensibly speak about them. This has

---

[9] Basic Law V states that the value-ranges of two functions f and g are the same iff $\forall x(f(x) = g(x))$.

[10] Second-order logic is logic that allows quantification over predicates as well as over individuals. First-order logic is logic that quantifies only over individuals. There is some debate over whether second-order logic really is logic or merely disguised set theory.

[11] See, for example, Boolos (1987, 1998), Burgess (2005), Hale and Wright (2001), Wright (1983), and Zalta (1999, 2000) for modern neo-logicist approaches. The classic original logicist treatises are Frege (1967, 1974), and Whitehead and Russell (1910, 1912, 1913).

ramifications for both the style of proof that is acceptable in mathematics and the domains of mathematical objects one can work with. Unless there is a procedure for delivering the mathematical objects in question, they are committed to the flames. All but the smallest, most well-behaved infinities are rejected. But most notable is that many proofs of classical mathematics are not valid by intuitionistic lights.

To understand why, think about the theorem of classical logic known as the *law of excluded middle*: for every proposition P, the disjunction of P and its negation, $(P \vee \neg P)$, is true.[12] This law is well motivated in cases where we may be ignorant of the facts of the matter, but where there *are* facts of the matter. For example, the exact depth of the Mariana Trench in the Pacific Ocean at its deepest point at exactly 12.00 noon GMT on 1 January 2011 is unknown, I take it. But there is a fact of the matter about the depth of this trench at this time. It was, for example, either greater than 11,000 m or it was not. Contrast this with cases where there is plausibly no fact of the matter. Many philosophers think that future contingent events are good examples of such indeterminacies. Take, for example, the height of the tallest building in the world at 12.00 noon GMT on 1 January 2031. According to the line of thought we're considering here, the height of this building is not merely unknown, the relevant facts about this building's height are not yet settled. The facts in question will be settled in 2031, but right now there is no fact of the matter about the height of this building. Accordingly, excluded middle is thought to fail here. It is not, for example, true that either this building is taller than 850 m or not.

Now consider mathematics, as understood by the intuitionists. For them, mathematics is all about the construction of mathematical objects and proofs concerning them. Let's focus on the proofs. Consider some mathematical statement S that is neither proven nor proven to be false. If one does not recognise some objective, external sense of truth, and instead takes proof to be all there is to it, excluded middle fails for S. In particular, excluded middle cannot be used in the process of proving S. Double-negation elimination also fails. After all, proving that there is no proof that there can't be a proof of S is not the same thing as having a proof of S. The rejection of double-negation elimination undermines an important form of

---

[12] Excluded middle should be carefully distinguished from its semantic counterpart, *bivalence*: every proposition is either true or false.

proof in classical mathematics known as *reductio ad absurdum*. This style of proof starts by assuming the negation of S, then proceeds to draw a contradiction from this assumption, thus concluding simply S.[13] In intuitionistic logic, this is all fine until the last step. According to the intuitionist, all that has been shown is $\neg\neg S$ and it is a further unjustified step to conclude S from this. Some other classical forms of proof are intuitionistically invalid. These include various existence proofs that show that some object must exist but do not deliver a construction of the object in question (e.g., the proof of the Tarski–Banach Theorem in section 9.1.1 is not intuitionistically valid). Intuitionism is thus a more radical philosophy of mathematics than the others we've seen so far, in that it demands a change in mathematical practice. It requires a new logic, with many traditional proofs of mathematical results no longer accepted.[14]

1.2 Charting a course to contemporary topics

The agenda for contemporary philosophy of mathematics was shaped by Paul Benacerraf in a couple of landmark papers. In the first of the papers (Benacerraf 1983a, originally published in 1965), Benacerraf outlines an underdetermination problem for the project of reducing all of mathematics to set theory. Such underdetermination or non-uniqueness problems had been around for some time, but Benacerraf's presentation was compelling, and its relevance to a popular position in philosophy of mathematics was firmly established. The second and third problems (Benacerraf 1983b, originally published in 1973) are presented as a challenge that any credible philosophy of mathematics must meet: (i) allow for a semantics that is uniform across both mathematical and non-mathematical discourse and (ii) provide a plausible epistemology for mathematics. As Benacerraf went on to show, it is difficult to satisfy both parts of this challenge simultaneously. Any philosophy of mathematics that meets one part of the challenge typically has serious difficulties meeting the other part.

There are two main camps in philosophy of mathematics and each has a serious problem with one or other of these challenges. Realist or

---

[13] See section 9.1.9 for an example of such a proof in mathematics.

[14] For more on intuitionism see Heyting (1971; 1983), Dummett (1983), and Brouwer (1983).

Platonist philosophies of mathematics[15] hold that at least some mathematics is objectively true and is about a realm of abstract mathematical entities. Mathematics is taken at face value and the semantics here is the same as elsewhere. Mathematical realism has no problem with the first of Benacerraf's challenges but, notoriously, has serious difficulties providing a plausible epistemology. Anti-realist positions, on the other hand, hold that there are no such abstract mathematical entities. The anti-realist thus has no epistemic problems, but these positions typically fall foul of the first of Benacerraf's challenges.

The three Benacerraf problems, along with a few others we'll encounter, are the rocks on which many philosophies of mathematics founder. The challenge is to chart a course past these difficulties to arrive at a credible philosophy of mathematics. So let's get better acquainted with the main obstacles.

1.2.1 Uniform semantics

The requirements for a uniform semantics is just that one should not give special semantic treatment to mathematical discourse. If a mathematical statement such as '$\sqrt{2}$ is irrational' is taken to be true, the semantics should be the same as for other true sentences such as 'Jupiter is a gas giant'. The latter is true by virtue of the existence of Jupiter and it having the property of being a large planet composed primarily of the gases hydrogen and helium. Under a uniform semantics, '$\sqrt{2}$ is irrational' is true by virtue of the existence of the number $\sqrt{2}$ and it having the property of not being expressible in the form a/b, where a and b are integers. The requirement of providing a uniform semantics leads very naturally from truth of mathematical statements to the existence of mathematical objects. Mathematical realism thus has a very natural answer to this challenge. It is anti-realism that has difficulties here. For example, if your view is that what makes '$\sqrt{2}$ is irrational' true is something about a social agreement to assent to such claims or to the existence of a proof of an appropriate kind, then the requirement for uniform semantics requires that you do the same for the sentence above about Jupiter. Either you give a deviant semantics across the board or you use the usual semantics in mathematics as well. Of course

---

[15] I will use the terms 'Platonism' and 'mathematical realism' interchangeably.

you may decide to treat the semantics of mathematics differently and violate the requirement for uniform semantics, but then you at least owe an explanation why mathematics comes in for such special treatment.

1.2.2 The epistemic problem

The epistemic problem is very simple: provide an account of how we come by mathematical knowledge. The problem was originally cast in terms of the causal theory of knowledge. This theory holds that for an agent A to know some proposition P, A must believe that P, and the fact that makes P true must cause A's belief that P. Thus construed, the epistemic challenge was to show how mathematical knowledge could be reconciled with the causal theory of knowledge. For Platonist accounts of mathematics, this was nearly impossible, for it would mean coming in causal contact with mathematical entities: the number 7, for instance, would need to cause my belief that 7 is prime. But surely numbers do not have causal powers. Indeed, it would seem that numbers are the wrong kind of thing to be causing anything, let alone beliefs. This leads many to be wary of, if not outright reject, Platonism.

But there are problems with the argument, thus construed. For a start, why should we accept the causal theory of knowledge? After all, this theory was formulated with empirical knowledge in mind and was not intended to deal with mathematical knowledge. It is question-begging to require the Platonist to provide a causal account of mathematical knowledge. If anything should be rejected here, it should be the causal theory of knowledge. In any case, the causal theory of knowledge did eventually fall from grace. The reasons for this were various, but its inability to account for mathematical knowledge was chief among its deficiencies. Still, many seem to think there's something to Benacerraf's challenge which survives the demise of the causal theory of knowledge. W. D. Hart puts the point thus:

> [I]t is a crime against the intellect to try to mask the problem of
> naturalizing the epistemology of mathematics with philosophical
> razzle-dazzle. Superficial worries about the intellectual hygiene of causal
> theories of knowledge are irrelevant to and misleading from this problem,
> for the problem is not so much about causality as about the very possibility
> of natural knowledge of abstract objects. (Hart 1977, pp. 125–6)

What is the worry about abstract objects? What is it about abstract objects that suggests that it's impossible to have knowledge of them? In my view, the most cogent post-causal-theory-of-knowledge version of this argument is due to Hartry Field. He captures the essence of the Benacerraf argument when he puts the point in terms of explaining the reliability of mathematical beliefs (emphasis in the original):

> Benacerraf's challenge – or at least, the challenge which his paper suggests to me – is to provide an account of the mechanisms that explain how our beliefs about these remote entities can so well reflect the facts about them. The idea is that *if it appears in principle impossible to explain this*, then that tends to *undermine* the belief in mathematical entities, *despite* whatever reasons we might have for believing in them. (Field 1989, p. 26)

Put slightly differently, the challenge is to account for the reliability of the inference from 'mathematicians believe that P' (where P is some proposition about some mathematical object(s)) to 'P', while making explicit the role that the mathematical entities play in this reliable process. Mathematical entities don't necessarily have to cause the mathematicians' beliefs, but mathematical entities need to be part of the story. Coming up with any such plausible story has proven to be the Achilles' heel of Platonism.

1.2.3 Underdetermination

The problem here is one of an embarrassment of riches. Many philosophers and mathematicians are inclined to identify various mathematical objects with some set-theoretic construction or other. So, for example, we can construct a set-theoretic counterpart of the natural numbers that does everything we want of the natural numbers. We then identify the natural numbers with this set-theoretic construction. If we take this identification seriously, we then think of the natural numbers as sets. (Or, equivalently, we claim that the natural numbers have been reduced to sets.) It turns out that there's not much in mathematics you can't construct out of sets. This leads some to suggest that all you need is sets. This is a very seductive line of thought, for it means that if we must believe in mathematical objects, we at least need only one kind of mathematical object, namely, sets. The rest come for free. This set-theoretic-reduction strategy is thus ontologically more parsimonious than some of the alternatives and gives mathematics a rather appealing unity.[16]

---

[16] Indeed, this is one reason why philosophers of mathematics tend to concentrate so much on set theory.

The problem is, however, that there are just too many ways to affect the set-theoretic constructions in question. Take the natural numbers as our example. We have the set of finite von Neumann ordinals:[17] $0 = \emptyset$; $1 = \{\emptyset\}$; $2 = \{\emptyset, \{\emptyset\}\}$; $3 = \{\emptyset, \{\emptyset\}, \{\emptyset, \{\emptyset\}\}\} \ldots$, where the successor function $S(x) = x \cup \{x\}$. We also have the set of finite Zermelo ordinals:[18] $0 = \emptyset$; $1 = \{\emptyset\}$; $2 = \{\{\emptyset\}\}$; $3 = \{\{\{\emptyset\}\}\} \ldots$, where the successor function is $S(x) = \{x\}$. There are many other set-theoretic models of the natural numbers, but these two are enough to raise a problem. The problem is that if the natural numbers are sets, we need to be able to say which sets they are. Is 3 $\{\emptyset, \{\emptyset\}, \{\emptyset, \{\emptyset\}\}\}$ or $\{\{\{\emptyset\}\}\}$? Benacerraf makes the point by telling a story of two children Ernie (for Ernst Zermelo) and Johnny (for John von Neumann) learning basic arithmetic set theoretically via the above constructions. Ernie learns that the von Neumann ordinals are the natural numbers, while Johnny learns that the Zermelo ordinals are the natural numbers. Each child does well with their arithmetic until they start asking questions such as: is $1 \in 3$? Ernie says 'yes', while Johnny says 'no'. Benacerraf's point is that if the natural numbers are sets, at most one of Johnny and Ernie is right (they might both be wrong – the natural numbers might be some entirely different set-theoretic construction).

What is interesting is that if we confine ourselves to purely arithmetic questions, Ernie and Johnny agree, so the fact that they are using different natural number systems doesn't matter for purposes of counting and the like. But the two set-theoretic constructions are clearly different, and the difference can be brought out via perfectly good mathematical questions. There is something odd about the questions needed to bring out the differences – questions such as 'is $1 \in 3$?' – but for someone who sees the natural numbers as sets, such questions are more natural than they may first appear.

The conclusion to be drawn from all this is that since we can't have more than one set of natural numbers, the natural numbers can't be sets at all.

---

[17] Named for the Hungarian-born mathematician John von Neumann (1903–57). These are *ordinal numbers*, meaning they are numbers representing order relations. These should be contrasted with *cardinal numbers*, which we will meet in the next chapter. The latter represent the size of a set. An example of an ordinal number is the 3 in 'Chapter 3 of this book'; it tells us about the position of that chapter in the book's running order. Contrast this with the cardinal number three in 'The Three Stooges': it tells us how many stooges there are.

[18] Named for the German mathematician Ernst Zermelo (1871–1953).

A very natural thought at this point is to focus on the structure in each case. Perhaps what is important is not the objects themselves but, rather, the structural relations between them. Perhaps the natural numbers are anything that has the right kind of structure – it could be the finite von Neumann ordinals, the finite Zermelo ordinals, or anything else with the desired structure – known as an *ω-sequence*.[19] This move away from objects to structures has a great deal to recommend it and we will return to the idea in Chapter 3. For now, we just note that if our philosophy of mathematics has it that numbers are sets, we need an answer to this Benacerraf underdetermination problem.

1.2.4 Other issues

There are, as you might expect, several other obstacles around which we must navigate on our way to a cogent philosophical account of mathematics. Some of these will arise in later chapters. For instance, many philosophers subscribe to some version of a doctrine known as *naturalism*. In its most general form, naturalism is a rejection of 'spooky', unscientific, or other-worldly accounts of the way the world is. One particularly influential way of spelling this out, due to the philosopher W. V. Quine (1908–2000), is in terms of the relationship between philosophy and science. According to Quine, philosophy is not prior to science, nor does philosophy take priority over science. Rather, philosophy is continuous with the scientific enterprise. He sees naturalism as the

> abandonment of the goal of a first philosophy. It sees natural science as an inquiry into reality, fallible and corrigible but not answerable to any supra-scientific tribunal, and not in need of any justification beyond observation and the hypothetico-deductive method. (Quine 1981a, p. 72)

This places constraints on the authority of philosophy. Philosophy must take the relevant science seriously and, in general, philosophy cannot overturn science on purely philosophical grounds. Quine seemed only to acknowledge empirical science, but we might follow Penelope Maddy (1997) here and extend this philosophical humility so that we also respect the methods and results of mathematics. Whether or not you want to call this

---

[19] See section 9.4 for a little about ω, the first infinite ordinal number.

'naturalism', it does seem right that philosophers should not be too quick to criticise mathematical methodology and results on philosophical grounds. This is not to suggest that philosophy has no role in understanding mathematics, or that philosophers should merely second the pronouncements of mathematicians. Philosophy does have a role here, but it is a subtle one, involving serious engagement with the relevant mathematics and science. The issue of naturalism and respecting mathematical methodology are recurring themes throughout the rest of this book.

Another constraint on a good philosophy of mathematics is that it needs to provide an account of mathematics in applications. As we saw earlier in this chapter, formalism had trouble accounting for mathematics in applications. After all, if mathematics is a game akin to chess, as the game formalists would have it, why is mathematics in such high demand in the formulation and interrogation of scientific theories? This line of thought has been developed into an argument for mathematical realism. In its modern guise, the argument is due to W. V. Quine (1953) and Hilary Putnam (1971). This *indispensability argument* will be considered in some detail in Chapter 3, along with some anti-realist responses to it in Chapter 4. But there are other issues involved in providing a complete account of mathematics in applications – mathematical realism isn't the answer to all philosophical problems concerning mathematics in applications. We will consider some of these further issues in subsequent chapters. In Chapter 5 we consider the role mathematics plays in both mathematical and scientific explanations, and in Chapter 6 we consider the question of whether there is something unreasonable and mysterious about the applicability of mathematics.

Finally, a couple of topics that don't usually see their way onto the mainstream philosophy of mathematics agenda: inconsistent mathematics and mathematical notation. Both of these topics have their roots in earlier positions. Recall that Russell's proof of an inconsistency in Frege's logicist programme prompted the search for both a consistent set theory to replace the inconsistent naïve set theory and a secure (consistent) foundation for mathematics. But the fact that mathematics did not go 'belly up' as soon as the contradiction was found is intriguing and prompts a number of questions about the success and limits of inconsistent mathematics. We pursue some of these questions in Chapter 7. The issues associated with mathematical notation also go back to one of the big isms – this time, formalism.

As we have already seen, and to its great credit, formalism treated mathematical notation very seriously. But since formalism lost support, the issues associated with notation and mathematical representation, more generally, were sidelined. This strikes me as a mistake, and Chapter 8 is a plea to put mathematical notation back on the agenda.

1.3 Planning for the trip

1.3.1 For the student

Philosophy of mathematics, when done right, should take you through some fascinating mathematical territory. The philosophy of mathematics is also interesting in its own right and it connects up with important issues elsewhere in philosophy. But what really sets it apart is the mathematics. In the interests of keeping the mathematics simple, and the topic accessible to a wide audience, many treatments of the philosophy of mathematics focus on simple, pedestrian arithmetic statements such as $7 + 5 = 12$, with perhaps a bit of set theory thrown in. That's a bit like trying to keep your trip to Europe simple by only visiting Glasgow. As interesting as Glasgow is, it is not representative of all of Europe. Travelling can be inconvenient, tiring, and sometimes downright hard work, but the rewards are great. If you want to see and learn about Europe, you need to do the work and venture beyond Glasgow. To hell with the pedestrian, live adventurously!

I suggest the same policy for the philosophy of mathematics. Mathematics is a rich and interesting field and, as a philosopher, you just wouldn't be doing it justice if you concentrated on pedestrian numerical statements and the basics of set theory. To venture beyond these, into number theory, real and complex analysis, topology, differential equations, and modern algebra requires work, but the pay-off is a more rounded philosophy of mathematics and a much richer experience. You'll become acquainted with some extraordinary mathematical results – some of which are among humanity's greatest achievements. It's definitely worth the effort.

Of course, in a book such as this I cannot give a full and proper treatment of all the mathematical topics I'd like to cover. I try to outline the basics of some of the more philosophically interesting mathematical results in terms a keen student of philosophy would find accessible. But this is not a mathematics text, so occasionally I will need to skate over details. In such cases I encourage the student to pursue the details independently. Indeed, all the mathematics covered in this book deserves deeper investigation than I can offer here. In short, I try to steer a course through some philosophically interesting mathematics in such a way that doesn't presuppose too much prior knowledge from the reader, but also doesn't 'dumb down' the mathematics. I assume that anyone interested in philosophy of mathematics is acquainted with, at least, high-school mathematics and perhaps has an introductory logic course under their belt. But anyone, with the aid of a good mathematics dictionary or resource book, should be able to follow the mathematics in this text. Sometimes a little side trip into the mathematics will be required, but that is as you would expect. Indeed, I hope this book prompts many such side trips into the mathematical material in question.

1.3.2 For the instructor

There are, of course, topics in the philosophy of mathematics not covered in this book, and others not much more than mentioned. For a start, there is very little about the history of philosophy of mathematics and the big movements of the first half of the twentieth century. This material is already covered in a number of readily accessible places. I don't have anything to add to what has been said (many times) elsewhere. My aim in this book is to bring students up to date with the contemporary philosophy of mathematics scene, and in a couple of instances introduce them to some relatively unexplored topics. But even when limiting my attention to the contemporary scene, I've had to make choices about what to cover. Unfortunately, some topics have been relegated to mere passing mentions and some omitted completely. This is not to suggest that topics given such short shrift are unimportant. It's just that I cannot do justice to all the issues currently receiving attention in contemporary philosophy of mathematics, so I picked some of my favourites and ran with those. Occasionally students are invited to consider some of the neglected topics via the discussion questions at the end of each chapter.

I've opted for relatively detailed treatments of the topics covered so that the student gets their teeth into at least some of the issues capturing the attention of contemporary philosophers of mathematics. An alternative would be to go for breadth, trying to say something about all the major topics. Such surveys can be found and certainly have an important place in teaching. But for my money, I'd rather get into the details of a topic than to survey a number of topics. This allows the student to *do* philosophy, rather than merely learn about it. For instance, in the chapter on nominalism (Chapter 4) I focus on just three strategies: Hartry Field's fictionalism, and the nominalist strategies suggested by Jody Azzouni and by Stephen Yablo. Although I think that these are three of the more important nominalist strategies going around, there are several other strategies deserving of attention. My thinking here is that if the student becomes reasonably familiar with these three, they will have no trouble following other nominalist strategies. Whereas, were we to take the superficial survey approach, I am not convinced that the student would acquire the required skills and knowledge to really engage with any of the positions in question. In any case, that's the thinking behind my choice of topics and decision to give some topics merely cursory treatment. If any of my omissions are too much for you to bear, it should be easy enough to present supplementary material at the relevant points in the course. Indeed, some of the suggestions for further reading at the end of each chapter are intended as such points of departure.

I include an epilogue, which presents a number of mathematical results, open questions, and interesting numbers – all in list form. Some of this material is covered elsewhere in the book, but some is not. In my view, budding philosophers of mathematics would do well to be familiar with the mathematics touched on in the epilogue, and there I have briefly indicated why. This chapter could be read at the end of a philosophy of mathematics course as a revision of the philosophical material, while extending the range of mathematical examples covered. It could also be integrated into a course, and, of course, the instructor is free to cherry-pick items from my lists and add their own, as they see fit. Some of the items in my lists might be chosen to be covered in greater detail by fleshing out both the relevant mathematics and the philosophical issues they give rise to.

Discussion questions

1. What are the primary objects of study of mathematics?
2. Why might formalism have trouble accounting for mathematics in applications?

3. What does it mean to say that mathematics is logic? Spell out the similarities and differences you take to exist between logic and mathematics.
4. Do you agree that the law of excluded middle fails for future contingents?
5. According to intuitionists, a mathematical theorem is true just in case there is an intuitionistically acceptable proof of the theorem in question. Does this mean that the truth of mathematical theorems can change with time? How plausible is this?
6. Why might the semantics of mathematics be different from the semantics of the rest of language?
7. Can you think of any other examples, besides mathematics, where we have knowledge of some proposition P, which is not derived from causal contact with whatever it is that makes P true?
8. Might we solve Benacerraf's underdetermination problem by insisting that one set-theoretic construction of the natural numbers is the correct one? Can you think of any good candidate constructions?

Recommended further reading

For further reading on formalism see von Neumann (1983) and Weir (2011). For good overviews of the logicist and neo-logicist programmes see MacBride (2003) and Zalta (2010). For more on intuitionism, see Brouwer (1983), Dummett (1983), and Heyting (1983). For a nice presentation of intuitionistic logic see Priest (2008). The classic Benacerraf papers are (1983a) and (1983b). These two are essential reading.

Benacerraf, P. 1983a. 'What Numbers Could Not Be', in P. Benacerraf and H. Putnam (eds.), *Philosophy of Mathematics: Selected Readings*, 2nd edn, Cambridge University Press, pp. 272–94.
    1983b. 'Mathematical Truth', in P. Benacerraf and H. Putnam (eds.), *Philosophy of Mathematics: Selected Readings*, 2nd edn, Cambridge University Press, pp. 403–20.
Brouwer, L. E. J. 1983. 'Intuitionism and Formalism', in P. Benacerraf and H. Putnam (eds.), *Philosophy of Mathematics: Selected Readings*, 2nd edn, Cambridge University Press, pp. 77–89.
Dummett, M. 1983. 'The Philosophical Basis of Intuitionism', in P. Benacerraf and H. Putnam (eds.), *Philosophy of Mathematics: Selected Readings*, 2nd edn, Cambridge University Press, pp. 97–129.

Heyting, A. 1983. 'The Intuitionist Foundations of Mathematics', in P. Benacerraf and H. Putnam (eds.), *Philosophy of Mathematics: Selected Readings*, 2nd edn, Cambridge University Press, pp. 52–60.

MacBride, F. 2003. 'Speaking with Shadows: A Study of Neo-Logicism', *British Journal for the Philosophy of Science*, 54: 103–63.

Neumann, J. von 1983. 'The Formalist Foundations of Mathematics', in P. Benacerraf and H. Putnam (eds.), *Philosophy of Mathematics: Selected Readings*, 2nd edn, Cambridge University Press, pp. 61–5.

Priest, G. 2008. *An Introduction to Non-Classical Logic: From If to Is*, 2nd edn, Cambridge University Press, Chapter 6.

Weir, A. 2011. 'Formalism in the Philosophy of Mathematics', in E. N. Zalta (ed.), *The Stanford Encyclopedia of Philosophy* (Spring 2011 edn), http://plato.stanford.edu/archives/spr2011/entries/formalism-mathematics/.

Zalta, E. N. 2010. 'Frege's Logic, Theorem, and Foundations for Arithmetic', in E. N. Zalta (ed.), *The Stanford Encyclopedia of Philosophy* (Fall 2010 edn), http://plato.stanford.edu/archives/fall2010/entries/frege-logic/.

Useful general resources

Now to some useful general resources which will be good to have access to for any course on the philosophy of mathematics. First, a few good overviews of the philosophy of mathematics. Brown (2008) is an excellent introduction to the subject, with a contemporary focus. Bostock (2009) and George and Velleman (2002) are historically-oriented introductions. Friend (2007) and Shapiro (2000) present good mixes of historical and contemporary issues.

There are various collections of key papers on the philosophy of mathematics. These will be useful in filling out the material in this text with some of the primary sources. Benacerraf and Putnam (1983) is a classic collection. It's a bit out of date, but most of the papers are still core reading, especially for the traditional philosophy of mathematics topics. Bueno and Linnebo (2009) is an interesting collection of papers with more of a focus on recent and evolving research areas in philosophy of mathematics. Hart (1996) is a short, affordable collection with many of the classic papers. Jacquette (2001) is a collection of papers arranged around a mixture of traditional and contemporary topics. Leng, Paseau, and Potter (2007) is a good collection of essays on mathematical epistemology, by philosophers of mathematics, mathematicians, and psychologists. Mancosu (2008a) is a good collection of papers on a topic of great contemporary interest. Shapiro (2005) is a terrific collection of new papers summarising the latest developments in most of the major areas of philosophy of mathematics, past and present. Tymoczko (1998) is an interesting collection of papers covering some less commonly explored topics.

Some students might be interested to pursue some of the history of the mathematics we encounter. Good sources here are Grattan-Guinness (2003, 2007) and Kline (1972). A good introduction to mathematical methods is Courant and Robbins (1978). Finally, a couple of good mathematical reference works. These can be useful for filling in holes in students' mathematical background and for pursuing some of the mathematics in greater depth. Borowski and Borwein (2002) is a good general-purpose mathematical dictionary; Gowers (2008) is an excellent general reference work on mathematics.

2   The limits of mathematics

One of the endlessly alluring aspects of mathematics is that its thorniest paradoxes have a way of blooming into beautiful theories.

Philip J. Davis (1923–)[1]

Most mathematics is concerned with proving theorems, developing new mathematical theories, and finding axioms for theories. But there are very important questions about the mathematical theories themselves. For example, it would be nice to know whether a particular mathematical theory is consistent. That is, we'd like to be able to prove that the mathematical theory in question will not deliver a contradiction, as naïve set theory did. We'd also like to know whether a mathematical theory is capable of answering any question thrown at it. That is, we'd like to be able to show that for any mathematical statement of the theory, either we can prove it or prove its negation. This is called completeness. The study of such higher-level questions about mathematics is known as *Metamathematics* and can be thought of as the mathematical study of mathematics. Not surprising, this is an area of great interest for philosophers, especially in light of a number of key results that place limitation on what mathematics can do. These results are intriguing (and often surprising) in their own right, but they are also supposed to have consequences for philosophy of mathematics and beyond – to areas such as philosophy of mind and metaphysics. In this chapter we consider some of these results and discuss their significance for philosophy.

---

[1] P. J. Davis, 'Number', *Scientific American*, 211 (September 1964): 51–9, at p. 55, Reproduced with kind permission of *Scientific American*.

2.1 The Löwenheim–Skolem Theorem

2.1.1 Background

The first limitative result we will look at is from set theory and is known as the Löwenheim–Skolem Theorem. To understand the significance of this theorem we first need to look at the work of one of the great nineteenth-century mathematicians, the Russian-born German Georg Cantor (1845–1918). Cantor did pioneering work in set theory, with his most celebrated result concerning the cardinality of sets. When considering finite sets, the notion of cardinality is straightforward: the cardinality of a set is the number of elements in the set, and here we can just count off the elements. Two finite sets have the same cardinality if and only if (iff) they have the same number of elements, otherwise one is larger than the other. But when we get to infinite sets, things get more interesting. Consider some of the well-known infinite systems of numbers. Intuitively, there are more integers than there are natural numbers; there are more rational numbers than there are integers; there are more real numbers than there are rational numbers; and there are more complex numbers than there are real numbers. But how do you count the numbers in question, given that in all cases there are infinitely many of them? How do we get a grip on the cardinality of infinite sets? Must we admit that all infinite sets are the same size and give up on the above intuitions?

Cantor gave us a very useful way forward: he proposed that two sets have the same cardinality iff the members of one set can be placed in one–one correspondence with the members of the other. This is clearly right for finite sets, and even seems trivial. The idea is that when we have an auditorium with no one standing and no spare seats, we can say that the number of people is the same as the number of seats – we don't do this by counting the people and counting the seats and noting that they have the same cardinality. Rather, we notice the one–one correspondence between chairs and people, and we exploit that correspondence.[2] So much for the

---

[2] Another related application of one–one correspondence is in the Pigeonhole Principle or Dirichlet's Box Principle. This principle tells us if we have m objects and n boxes to place them in, if $m > n$ then at least one box will contain more than one object. It is named after the French mathematician J. P. G. L. Dirichlet (1805–59). Despite sounding trivial, this principle has a number of non-trivial applications. But to give a trivial application,

finite case. In the infinite case, Cantor's definition of 'same cardinality' is anything but trivial.

It turns out that we can place the natural numbers and the integers in one–one correspondence by the following trick: count the integers off as follows, 0, 1, −1, 2, −2, 3, −3, ... This mapping and Cantor's definition establishes (surprisingly) that the cardinality of the integers and the natural numbers is the same, despite the fact that the natural numbers form a proper subset of the integers. Similar tricks can be exploited to show that the rational numbers can also be placed in one–one correspondence with the natural numbers. The natural numbers, the integers, and the rationals all have the same cardinality, designated $\aleph_0$ and are said to be *countable*.[3] $\aleph_0$ is the smallest infinite cardinal number. We see that being a proper subset does not entail lower cardinality – that is only guaranteed for finite sets. This already leads to some puzzling results, such as Hilbert's Hotel. This is a hotel with infinitely many rooms and with each of the rooms occupied. But when a new guest arrives, there's no need to turn her away, merely because the hotel is full. A simple reorganisation of guests is required, so that the guest in room 1 is moved to room 2, the guest in room 2 is moved to room 3, and so on. This series of moves accommodates all the existing guests and leaves room 1 vacant for the new arrival. This is strange but not paradoxical. It's just one of the peculiarities of infinite sets – peculiarities that run counter to our usual experiences and intuitions formed by consideration of finite cases.

Next we introduce the notion of the power set. The power set, $\mathcal{P}(A)$, of a set A is the set of all subsets of A. For example, the set $B = \{0, 1, 2\}$ has the following subsets: $\emptyset$ (the empty set), $\{0\}$, $\{1\}$, $\{2\}$, $\{0, 1\}$, $\{0, 2\}$, $\{1, 2\}$, $\{0, 1, 2\}$. The power set is thus $\mathcal{P}(B) = \{\emptyset, \{0\}, \{1\}, \{2\}, \{0, 1\}, \{0, 2\}, \{1, 2\}, \{0, 1, 2\}\}$. Notice that the cardinality of $\mathcal{P}(B)$ (denoted by $|\mathcal{P}(B)|$) is greater than the cardinality of B. The former has eight elements while the latter has three elements. This, it turns out, is very important. And notice that $|\mathcal{P}(B)| = 2^{|B|}$. This is also important.

Now we get to Cantor's celebrated theorem. This theorem tells us that the cardinality of the power set of a set is strictly greater than the cardinality of

---

it tells us that if you have more than seven friends, at least two of your friends were born on the same day of the week.

[3] A set is said to be *countable* (or denumerable) if its cardinality is finite or $\aleph_0$, that is, the set in question can be placed in one–one correspondence with some subset of the natural numbers. Obviously a countably infinite set has cardinality $\aleph_0$.

the original set. The proof is fairly straightforward and it will be instructive to run through it.

Theorem 1 (Cantor). *For any set A, $|\mathcal{P}(A)| > |A|$.*

Proof. If $A = \emptyset$ then its cardinality $|A| = 0$ and $|\mathcal{P}(A)| = 1$. So assume that A is non-empty. Since there are singleton sets $\{a\}$ for every $a \in A$, $|\mathcal{P}(A)| \geq |A|$. Now assume that $|\mathcal{P}(A)| = |A|$. That is, there is a 1–1 function $f : A \to \mathcal{P}(A)$. Let $z = \{a \in A : a \notin f(a)\} \in \mathcal{P}(A)$. So since f is 1–1, there exists a $w \in A$ such that $f(w) = z$. But $w \in f(w)$ iff $w \in z$ iff $w \notin f(w)$, and we have a contradiction. There is thus no such f and $|\mathcal{P}(A)| > |A|$. $\square$

Note that there is nothing in the proof that presupposes anything about the cardinality of A. In particular, when A is an infinite set we find that the power set of A is a larger infinite set. So, for example, the cardinality of the power set of the natural numbers is larger than the cardinality of the natural numbers. Indeed, it can be shown that the cardinality of a power set of some set A is $2^{|A|}$, as the example in the previous paragraph illustrated. The upshot of Cantor's Theorem is that there are uncountable (or non-denumerable) sets. These are sets that cannot be placed in one–one correspondence with the natural numbers. Such sets have cardinalities greater than $\aleph_0$. Indeed, by taking power sets of power sets of an initial infinite set, we can generate infinitely many infinite sets, each with strictly greater cardinality than its predecessor. Returning to our example of the various number systems, we find that although the natural numbers, the integers, and the rationals all have the same cardinality, the set of real numbers has a cardinality larger than these. The real numbers are uncountable and can be shown to have cardinality $2^{\aleph_0}$. The complex numbers, too, have this as their cardinality, as does $\mathbb{R}^n$ (an n-dimensional real space) for any finite n.

Cantor's Theorem is a fascinating result in its own right, establishing that if there is one infinite set, there are larger and larger infinities, and apparently no end to the size of these infinities. This was disturbing news at the time Cantor was working. It prompted some to eschew infinities altogether. But let's set aside such a radical response to Cantor's Theorem. For present purposes, what is important is that Cantor's Theorem establishes the existence of non-denumerable sets. Indeed, it is a theorem of set theory that there are such sets. This brings us to the Löwenheim–Skolem Theorem and the associated paradox.

2.1.2 Skolem's paradox

In 1915 the German mathematician Leopold Löwenheim (1878–1957) proved a remarkable result. He proved that if a first-order sentence has a model at all, it has a countable model.[4] In 1922 the Norwegian mathematician Thoralf Skolem (1887–1963) generalised this result to systems of first-order sentences.[5] What is remarkable about these results is that they appear to fly in the face of Cantor's Theorem. The Löwenheim–Skolem Theorem seems to be telling us that we do not need to entertain infinities beyond the countable. In particular, it seems to be telling us that there are countable models of the real numbers and of set theory itself. This apparent conflict between Cantor's Theorem and the Löwenheim–Skolem Theorem became known as *Skolem's paradox*.

The problem can be reposed in terms of non-uniqueness of the domains in question. A set of axioms is said to be *categorical* if all models of the axioms in question are isomorphic. For example, take an exercise from elementary logic to find a model of the Dean Martin sentence 'Everybody loves somebody (or other)': $\forall x \exists y Lxy$. This has models of all finite cardinalities as well as a countable infinite model. For example, a tragic love triangle model has a domain of three objects $\{1, 2, 3\}$ and the extension of $L = \{(1, 2), (2, 3), (3, 1)\}$. That is, we have 1 loving 2, 2 loving 3, and 3 loving 1. An infinite model has a domain $\mathbb{N}$, with the extension of $L = \{(1, 2), (2, 3), (3, 4), (4, 5), ...\}$. Thus $\forall x \exists y Lxy$, if taken to be a single-axiom theory, is not categorical: the model of cardinality 3 is not isomorphic with the model of cardinality $\aleph_0$. But, of course there is no issue here; there's no reason to expect such a single-axiom system to be categorical in the first place.

---

[4] A model of a set of sentences is an interpretation of the sentences in question, according to which they are all true. The interpretation consists of a domain over which the quantifiers range, and a function that assigns extensions to the names and predicates in the sentences in question. A model can be thought of as a set-theoretic construction of the theory or system of sentences in question, and the cardinality of the model is just the cardinality of the domain over which the quantifiers range. A toy example is given below of models of the Dean Martin sentence.

[5] This is the so-called *downward Löwenheim–Skolem Theorem*. There is also an *upward Löwenheim–Skolem Theorem* that states that if a system of first-order sentences has a model, it has models of higher cardinalities.

The axioms for familiar mathematical structures, however, are different; such axioms are supposed to be capturing all that is important about the mathematical structures in question. But what the Löwenheim–Skolem Theorem tells us is that any first-order formulation of our mathematical structures will allow non-isomorphic models, if they have models at all. Some of these models (e.g., a countable model of the real numbers) are easily recognised to be unintended, but still they are (in some sense, legitimate) models of the axioms in question. Which model is the intended model is not settled by the mathematical formulation alone; our first-order formulations are not categorical.

Despite the hint of paradox here, there is in fact no paradox. The generally agreed-upon solution to the apparent paradox is that although under some interpretations of the mathematical terms in question (*set membership*, *successor*, *subset*, and the like) there will be uncountable models, under *different* interpretations of the terms in question there will be countable models. What is crucial is the failure of the theory to absolutely fix the reference of the mathematical terms. Axiomatic systems such as those under consideration give rise to a kind of relativity. But with this relativity in place, we find that the alleged paradox turns on an equivocation: we are supposed to conclude that the mathematical theory in question has both countable and uncountable models under *the very same interpretation* of the mathematical terms in question. But that is a mistake.

What we have is a surprising result that places limits on what we can expect from our axiomatic mathematical theories. But some have suggested that the kind of relativity one needs to embrace in order to escape the apparent difficulties presented by the Löwenheim–Skolem Theorem has broader philosophical consequences. Hilary Putnam (1926–) argues that the Löwenheim–Skolem Theorem undermines common-sense realism, not just about mathematics but elsewhere as well (Putnam 1980). The idea, very roughly, is that if we were to formulate our best scientific theories in a first-order language, we'd find the same relativity. We find that there is no fact of the matter about the reference of our theoretical terms. Moreover, the indeterminacy in question undermines any confidence we have in the ontology of our physical and even common-sense theories. This, in turn, invites a turn to anti-realism. Not surprisingly, Putnam's argument has been very influential in metaphysics and threatens to undermine common-sense realist positions about scientific theories.

2.2 Gödel's Incompleteness Theorems

2.2.1 Gödel's results

Perhaps the most famous limitative results in mathematics are due to the Austrian logician Kurt Gödel (1906–78). Gödel proved a couple of ground-breaking incompleteness results in the 1930s. The first of these, known as Gödel's *First Incompleteness Theorem*, says that any consistent axiomatic system, rich enough to be of mathematical interest, will have recognisably true sentences that are not provable in the system in question. In order to prove this result, Gödel developed an ingenious encoding system so that he could assign a number to every well-formed sentence of the language in question – the sentence's *Gödel number* – and with this resource we can formulate within mathematics a sentence that says of itself that it is not provable. This sentence is often referred to as *the Gödel sentence* of the system in question. Consider the options here. If, despite what the sentence says, it *is* provable, then we would have a derivation of a false sentence. So assuming that the system in question is consistent, we must not be able to prove the sentence in question. But since that is what the sentence claims, we have an example of a true sentence that is not provable in the mathematical system in question. There are what we might think of as 'blind spots' in mathematics:[6] true sentences that cannot be derived in the system (so long as the system in question is consistent). Or to state this in slightly different terms, we need to choose between consistency and completeness. Most take this choice to be clear-cut and choose consistency. But there is something to be said for inconsistent systems here. For a start, they can be complete.[7]

Gödel's *Second Incompleteness Theorem* states that any consistent system of sufficient complexity to be of mathematical interest cannot prove its own consistency. The First Incompleteness Theorem reveals a blind spot in any (consistent) mathematical system, a somewhat artificial sentence that is recognisably true but not provable. The Second Incompleteness Theorem uses this blind spot to deliver a more substantial blind spot. The sketch of the proof of the Second Incompleteness result is as follows. Consider the true sentence: 'If the theory in question is consistent, then the Gödel sentence is not provable.' This is equivalent to 'If the theory in question

---

[6] I borrow the phrase 'blind spot' in this context from Sorensen (1988).

[7] We will return to the issue of inconsistent mathematics in Chapter 7.

is consistent, then the Gödel sentence'. So any proof of the consistency of the theory in question could be used to prove the Gödel sentence. But the First Incompleteness Theorem tells us that no consistent theory can prove its own Gödel sentence. So no consistent theory can prove its own consistency.

2.2.2 Philosophical significance of Gödel's incompleteness results

These results are philosophically interesting for a number of reasons. They are interesting historically because they proved to be a major stumbling block for (and perhaps even marked the end of) formalism as a viable philosophy of mathematics. Recall that formalism is the position that regards mathematics as being concerned with the manipulation of symbols on pieces of paper, whiteboards, and the like, and nothing more. The most well-developed version of this position is usually attributed to the German mathematician David Hilbert (1862–1943).[8] Hilbert had a sophisticated philosophy of mathematics, which (according to the common interpretation, at least) saw finite mathematics as being about the symbols and the manipulation of these symbols. He called this finite fragment of mathematics 'real'. This mathematics needed to be axiomatised and thus put on a secure basis.

Infinite mathematics had a different status and was called 'ideal' by Hilbert; it was there to help simplify and round out the finite mathematics but was not on the same footing as the finite mathematics. But although Hilbert treated infinite mathematics as a second-class citizen in his mathematical society, he would not entertain abandoning it in favour of finite methods, as others had suggested. Famously, Hilbert defended Cantor's vision of infinite mathematics: 'No one shall drive us out of the paradise which Cantor has created for us' (Hilbert 1983, p. 191). But in order to justify the use of infinite mathematics, Hilbert required a proof of the consistency of mathematics. Moreover, this proof needed to be completed by the (according to his lights) more secure finitary methods. It is generally thought that Gödel's incompleteness results (especially the second one) put an end to Hilbert's programme: Gödel showed that the consistency proof Hilbert required was possible only if mathematics was inconsistent.

---

[8] See Franks (2009), Hallett (1990), and Sieg (1999) for more nuanced interpretations of Hilbert's philosophy of mathematics than I can provide here.

Another interesting philosophical application of Gödel's results is due to the philosopher J. R. Lucas (1929–) and was later taken up and developed further by the physicist Roger Penrose (1931–). Lucas starts by noting that the blind spots of mathematics, such as the Gödel sentence, are not blind spots for humans. After all, by mere inspection we can see the truth of the Gödel sentence, but, on pain of inconsistency, mathematics cannot deliver the truth of the Gödel sentence. This, Lucas suggests, marks an important difference between purely formal systems and the human mind. To put the point in slogan form: the mind is not a machine. But this slogan massively overstates the case. For a start, there are machines other than the classical mathematical deduction systems that Gödel's theorem is concerned with. Perhaps the human mind is not a machine like this, but whoever thought that it was? Moreover, we should not forget the role consistency plays in Gödel's results. After all, it is only consistent systems that have the blind spots in question. Nothing rules out the human mind being a kind of inconsistent machine.

It is worth noting that what Gödel's First Incompleteness Theorem tells us is that a consistent formal system of the kind in question (including a suitably programmed computer) will have a blind spot – its Gödel sentence. But this is not to say that another system or machine will not be able to derive the first system's Gödel sentence. But Gödel's First Incompleteness Theorem then guarantees that the second system will have a blind spot of its own – its own Gödel sentence – and so on for as many systems/machines as you care to contemplate. Can the human mind really recognise the truth of all these Gödel sentences? If so, why can't the human mind just be a kind of open-ended machine that can always switch to another program that does not have the blind spot under consideration? There are many questions to be asked and responses to be made to the intriguing Lucas and Penrose arguments. The accepted wisdom (with which I concur) is that the Lucas–Penrose arguments fail. But putting your finger on exactly why they fail is still interesting and has helped in advancing our understanding of the relationship between minds and machines.

For me, however, the importance of Gödel's results does not lie in either of the applications just mentioned. The results, like any good pieces of mathematics, are interesting in their own right. There is no need for further philosophical theses to be defended on the back of Gödel's results. The results were surprising at the time of their first publication and even today they strike many as a little disturbing. Gödel's Incompleteness Theorems tell us something very profound about formal systems and their limits. I'm not claiming that there are no interesting applications of the Incompleteness Theorems outside of logic and mathematics, just that no further applications are required to make the theorems philosophically interesting. The existence of blind spots in consistent mathematical systems is surely food enough for thought.

2.3 Independent questions

2.3.1 The continuum hypothesis

An independent question of a theory is one which cannot be answered either positively or negatively. This indeterminacy is not merely epistemic, either. It's not that we don't currently know the answer to the question under consideration. Rather, the question itself is left open by the relevant mathematical theory. There are many such questions in mathematics, but the most famous example is the question of the size of the continuum: what is the cardinality of the set of real numbers?

We have already seen that the cardinality of the real numbers is infinite and is larger than the natural numbers. Moreover, it can be shown that the cardinality of the real numbers is that of the cardinality of the power set of the natural numbers. So we know that the cardinality in question is $2^{\aleph_0}$. So far, so good, but a problem arises when we ask where this cardinality lies in relation to the other infinite cardinalities. If we denote the first non-denumerable infinity $\aleph_1$, the next $\aleph_2$, and so on, we want to know which of these cardinalities is $2^{\aleph_0}$. A very tempting answer is that $2^{\aleph_0} = \aleph_1$; the continuum is the smallest non-denumerable infinity. This suggestion is known as the *continuum hypothesis* and has a proud history involving some of the greatest mathematicians of modern times.

It was Cantor himself who suggested the continuum hypothesis as the somewhat speculative answer to the question of the size of the continuum. Then, in a very famous address delivered by David Hilbert to the International Congress of Mathematicians in Paris in 1900, the 10 most pressing open problems in mathematics were outlined.[9] This list was amazingly

---

[9] A couple of years later, when published, the list was extended to 23 open problems. These are known as 'the Hilbert Problems', and there is great prestige attached to solving any one of these problems.

influential in shaping twentieth-century mathematics. Solving the continuum hypothesis was the first item on Hilbert's list. In 1940 Gödel proved that the continuum hypothesis was consistent with standard set theory (so-called Zermelo–Fraenkel set theory with the axiom of choice or ZFC). But being consistent with standard set theory and being provable in standard set theory are obviously not the same thing, and the continuum hypothesis resisted proof. Then, in 1963, the US mathematician Paul Cohen (1934–2007) showed why it resisted proof. He proved that the negation of the continuum hypothesis is also consistent with ZFC. With Gödel's 1940 result in hand, Cohen's result implies that the continuum hypothesis is independent of standard set theory.

Here we have an important and very natural question about mathematics – what is the cardinality of the real numbers? – that cannot be answered by our best theory. There are a couple of options here. One is to suggest that the relevant theory needs to be enriched in order to make it capable of delivering an answer to this question. Another option is to accept that there are such open questions in mathematics and give up the idea that all interesting questions in mathematics are answerable. There is something to be said for each of these options and we find ourselves embroiled in a realism–anti-realism debate about mathematics.

2.3.2 A realist response

One natural line of thought is to suggest that if a theory doesn't tell us all we want to know, we need a better theory. Take an example from elsewhere in science. If our best theory of space and time doesn't tell us what happens in accelerated frames of reference (e.g., special relativity works only with unaccelerated frames), we need a better theory (e.g., general relativity, which works with both accelerated and unaccelerated frames). Special relativity's silence on what goes on in accelerated frames is not seen as anything particularly mysterious about accelerated frames; it's just that special relativity is at best only part of the story. There are facts of the matter about the goings-on in accelerated frames, and our full theory of space and time will deliver those facts. Even before general relativity was developed it was clear that this was the right attitude to have towards accelerated frames. From a certain realist perspective, at least, this is a very natural thought whenever one encounters a theory that is silent on an issue of interest. The silence is seen as a limitation of the theory in question, and not an indeterminacy in nature.

Borrowing this train of thought and applying it to our independent questions in set theory, we see the independence of the continuum hypothesis from ZFC as an indication that ZFC is not the full story; we need to enrich our set theory so that it will give us answers to fundamental questions such as the cardinality of the real numbers. After all, surely, $2^{\aleph_0} = \aleph_1$ or $2^{\aleph_0} \neq \aleph_1$.[10] Since ZFC cannot tell us which, we should replace or supplement ZFC with further axioms so that we do get an answer to the continuum hypothesis. A great deal of work in modern set theory can be understood in the light of this line of thinking: proposing new axioms that will settle some of the more important independent questions in ZFC.[11]

2.3.3 An anti-realist response

Another line of thought has it that the independent questions are revealing something important about the mathematical realm itself, rather than mere limitations of our mathematical theories. To motivate this line of thought, consider an open question in a work of fiction. For example, in the 1960 movie of H. G. Wells's *The Time Machine* there is nothing more to the story than what is said and what appears on the screen (and perhaps the logical and natural implications of what is said and what appears). There is no fact of the matter about details omitted from the story. So when the time traveller sets off for the future to help a fledgling society, he takes with him three books. What were the three books? Well, that's (quite deliberately) not part of the story, so it would seem that there's no fact of the matter about what the three books were. It is not the case that either the time traveller took or did not take John Stuart Mill's *Utilitarianism* with him. The question of the titles of the three books is indeterminate in the story, but we may choose to enrich the story as we wish, but none of these enrichments should be thought of as *the correct* story, for the simple reason that there is nothing more to the correct story than what appears in the movie, and the movie is silent on the issue at hand. It seems we might well explore

---

[10] Always be suspicious when a philosopher says 'surely'.

[11] See Maddy (1997) for a very accessible account of some of the new axioms entertained by modern set theorists.

extensions of the story in which Mill's *Utilitarianism* was taken by the time traveller and extensions of the story where that particular book was left on the shelf. The difference here is that while it might be fun to explore such extensions, we should take a different attitude towards these extensions. There is a sense in which any consistent extension is allowable but none is the correct one. Although some extensions might be more interesting than others.

Now apply this line of thought to our independent questions of set theory. We might see these independent questions as genuinely independent in the way that the titles of the three books in the movie of *The Time Machine* are. Just as in fiction, we are free to explore extensions – extending ZFC by, for example, adding new axioms – but there is a lack of factualness about these extensions. In some extensions the continuum hypothesis is true and in others it is not, but the extensions themselves, at least according to the line of thought I'm suggesting here, are all on a par. We might even take a pluralist attitude towards the extensions and allow all consistent extensions. But as in the fiction, this does not mean that anything goes. Some extensions will be more interesting than others, and these will attract more attention.[12] This pluralist attitude towards set theory also seems to sit well with the current mathematical practice, where the attitude of those proposing new axioms may well be thought to be in this pluralist spirit.[13]

Discussion questions

1. Explain why Hilbert's Hotel is not paradoxical. (Notice how the set-up of this puzzle relies on Cantor's account of cardinality.)
2. Consider the set of all cardinal numbers. What is the cardinality of this set? (This is a version of Cantor's cardinality paradox.) Does this puzzle

---

[12] For example, extensions of the story in which the time traveller takes important books available to him on political theory, such as John Locke's *Two Treatises of Government*, Thomas Hobbes's *Leviathan*, and Mill's *Utilitarianism*, are interesting. Whereas extensions in which the time traveller takes a few cheesy romance novels are much less interesting.

[13] Again, see Maddy (1997) for more on the debates in set theory over new axiom candidates. She, in fact, argues that mathematical practice does not take sides here and that there is no way of settling the realism–anti-realism question in mathematics.

raise problems for Cantor's notion of cardinality, as applied to infinite sets?

3. Spell out, in detail, the Putnam argument against common-sense realism. Do you think that this argument works? How would you defend realism against this line of attack?

4. If the mind is a kind of machine, do Gödel's results at least place restrictions on the kind of machine it is?

5. What is to be gained, if anything, by taking the less-travelled path and embracing inconsistent but complete mathematical systems?

6. Can you think of other arenas where it seems reasonable to accept that there are no facts of the matter? Think about the logic of such domains. Can an argument be made for the failure of the law of excluded middle in such domains?

7. Do you think that independent questions such as the question of the size of the continuum undermine the plausibility of mathematical realism?

Recommended further reading

For more on the the Löwenheim–Skolem Theorem, see Bays (2009) and Skolem (1922). For Putnam's application of the theorem to metaphysics see his classic paper (1980). There is a great deal of subsequent literature on the Putnam argument and the interested student should have no trouble tracking down many good articles on the topic. On Gödel's theorem there is a huge literature. A good place to start is at the beginning with Gödel (1992), which is a reproduction of his original article. Other good sources are Dawson (2005), Kennedy (2010) and Smith (2007). On the related issues of Hilbert's formalist programme and the infinite, see Hilbert's classic piece (1983) and Zach (2009). For applications of Gödel's Incompleteness Theorems to philosophy of mind, see Lucas (1961) and Penrose (1989). A very rich and entertaining book on this and related topics is Hofstadter (1979). For more on independent questions in mathematics and the extensions of set theory that aim to resolve them, see Gödel (1983) and Maddy (1997).

Bays, T. 2009. 'Skolem's Paradox', in E. N. Zalta (ed.), *The Stanford Encyclopedia of Philosophy* (Spring 2009 edn), http://plato.stanford.edu/archives/spr2009/entries/paradox-skolem/.

Dawson, J. W., Jr. 2005. *Logical Dilemmas: The Life and Work of Kurt Gödel*, Natick, MA: A. K. Peters Publishers.

Gödel, K. 1983. 'What Is Cantor's Continuum Problem?', reprinted (revised and expanded) in P. Benacerraf and H. Putnam (eds.), *Philosophy of Mathematics: Selected Readings*, 2nd edn, Cambridge University Press, pp. 470–85.

1992. *On Formally Undecidable Propositions of Principia Mathematica and Related Systems*, New York: Dover.

Hilbert, D. 1983. 'On the Infinite', in P. Benacerraf and H. Putnam (eds.), *Philosophy of Mathematics: Selected Readings*, 2nd edn, Cambridge University Press, pp. 183–201.

Hofstadter, D. 1979. *Gödel, Escher, Bach: An Eternal Golden Braid*, New York: Basic Books.

Kennedy, J. 2010. 'Kurt Gödel', in E. N. Zalta (ed.), *The Stanford Encyclopedia of Philosophy* (Fall 2010 edn), http://plato.stanford.edu/archives/fall2010/entries/goedel/.

Lucas, J. R. 1961. 'Minds, Machines and Gödel', *Philosophy*, 36: 112–27.

Maddy, P. 1997. *Naturalism in Mathematics*, Oxford: Clarendon Press.

Penrose, R. 1989. *The Emperor's New Mind: Concerning Computers, Minds, and the Laws of Physics*, Oxford University Press.

Putnam, H. 1980. 'Models and Reality', *Journal of Symbolic Logic*, 45(3): 464–82.

Skolem, T. 1922. 'Some Remarks on Axiomitized Set Theory', in J. van Heijenoort (ed.), *From Frege to Gödel: A Source Book in Mathematical Logic 1879–1931*, Cambridge, MA: Harvard University Press, 1967, pp. 290–301.

Smith, P. 2007. *An Introduction to Gödel's Theorems*, Cambridge University Press.

Zach, R. 2009. 'Hilbert's Program', in E. N. Zalta (ed.), *The Stanford Encyclopedia of Philosophy* (Spring 2009 edn), http://plato.stanford.edu/archives/spr2009/entries/hilbert-program/.

3 Plato's heaven

The laws of mathematics are not merely human inventions or creations. They simply 'are'; they exist quite independently of the human intellect.

M. C. Escher (1898–1972)[1]

Mathematical realism or Platonism is the philosophical position that mathematical statements such as 'there are infinitely many prime numbers' are true and that these statements are true by virtue of the existence of mathematical objects – prime numbers, in this case. That all seems fine until you think about the nature of the objects being posited. Where are these numbers? What are they like? How do we know about them? What about all the other mathematical objects: sets, functions, Hilbert spaces, and the like? Do all these exist as well? Are all mathematical objects made up of the same basic ingredients – sets, perhaps – or are they each a distinct kind of thing? Are these mathematical objects abstract or do they have causal powers and space-time locations? In any case, what is their relationship to the physical world? And most difficult of all: if mathematical knowledge is knowledge of these mathematical entities how do we come by such knowledge? Negotiating a set of answers to these questions, unsurprisingly, leads to a variety of different realist positions. In this chapter we will very briefly consider a few of the realist positions on offer, before looking in more detail at an influential argument for mathematical realism.

3.1 A menagerie of realisms

There are two realist theses worth distinguishing in relation to mathematics. The first is the objectivity of mathematics. This is the thesis that

---

[1] B. Ernst, *The Magic Mirror of M. C. Escher*, Cologne: Taschen, 1978, p. 35. Reproduced with kind permission of the estate of M. C. Escher. © The M. C. Escher Company, www.mcescher.com.

mathematical statements are objectively true. They are not made true by sociological conventions or by the attitudes or beliefs of mathematicians. They are not 'true in a sense' or 'true from a particular cultural perspective' or anything of the kind. Mathematical statements are true, and they are true independently of our beliefs and attitudes towards them. That's mathematical objectivity. Some are of the view that this is all it takes to be a realist about mathematics. But notice that there's nothing in this about there being any mathematical objects. As Hilary Putnam put it:

> The question of realism, as Kreisel long ago put it, is the question of the objectivity of mathematics and not the question of the existence of mathematical objects. (Putnam 1979, p. 70)

Others disagree, and take mathematical realism to be the thesis that some mathematical statements are objectively true and that they are made true by the existence of mathematical objects. For example, take the statement 'there is an even prime'. If this is taken to be objectively true, how could it be so unless there is an even prime. It seems a very quick path from objective truth to objects. In any case, here we have our first divide among realists: those who see the realist thesis as being only about objectivity as opposed to those who see it as involving mathematical objects. But that's not the end of the internal disputes.

The next point of disagreement concerns the nature of mathematical objects. The traditional view, which comes down to us from Plato (429–347 BCE), is that mathematical entities exist but that they are unlike physical objects. Mathematical objects are abstract entities – objects without causal powers and lacking space-time locations. Driven by epistemological concerns about how we might get in touch with such ghostly entities, we find mathematical realists offering a variety of epistemic accounts of how access is secured. One of the more radical proposals is to reject the abstractness of mathematical entities. According to this line of thought, mathematical entities are physical. On some accounts it is even claimed that we can see them. As Penelope Maddy (1990) once argued, every time you look in the refrigerator and see a dozen eggs you are seeing the set of 12 eggs. You are thus face to face with a mathematical object, namely a set.[2] There are debates about whether mathematical entities exist necessarily

---

[2] Bigelow (1988) also argues for a physicalist account of mathematics.

or only contingently, and whether they are sets or something else. These are all debates which hold considerable interest, but let me say just a little about a couple of particular positions in the philosophy of mathematics that help bring out how some of the disagreements above get their bite.

3.1.1 Full-blooded Platonism

For the sake of argument, let's accept that there are some mathematical entities. Now consider the question of which mathematical objects there are. If you're a realist of this bent you take it as clear that there are numbers, sets functions, and the like. But how far do you go? Do you accept the reality of everything dreamt of by mathematicians? Or do you prefer your realism a little less extravagant and accept only those parts of mathematics needed for some practical purpose? Is there some middle ground? Let's consider the extravagant option first.

According to *full-blooded Platonism* every consistent mathematical theory truly describes some part of the mathematical universe. This view delivers a very rich ontology indeed: every mathematical object that could exist, does exist. But this is not excess for the sake of excess. There is good reason for subscribing to such an inflated ontology. The idea is straightforward.

A defender of such a full-blooded or plenitudinous Platonism, Mark Balaguer, points out that it would be utterly mysterious were someone to have true beliefs about the day-to-day events in a remote village in Nepal, without a reliable mechanism explaining the correlation between the belief formation and the events in the village in question. But traditional Platonism finds itself in something very much like this situation: we are supposed to have true beliefs about an abstract realm of mathematical entities with which we can't have causal contact. Balaguer goes on to point out that:

> [I]f all possible Nepalese villages existed, then I *could* have knowledge of these villages, even without any access to them. To attain such knowledge, I would merely have to dream up a possible Nepalese village. For on the assumption that all possible Nepalese villages exist, it would follow that the village I have imagined exists and that my beliefs about this village correspond to the facts about it. (Balaguer 1998, p. 49)

This, of course, is hardly a tenable epistemology for Nepalese villages. But Balaguer's point is just that on the assumption that all possible Nepalese villages exist, there is no mystery about how we can have knowledge without causal contact.

Now return to the epistemic woes of Platonism. If every mathematical object that could exist, does exist, our mathematical beliefs couldn't miss – they would latch on to some part of the mathematical realm. In short, our beliefs about the mathematical objects of a consistent theory constitute knowledge of those objects.

The core idea here is very clever: you can overcome an epistemic access problem by inflating ontology to the limit. Moreover, this view has a ready reply to many underdetermination problems. The idea is to embrace non-uniqueness. Full-blooded Platonism is *not* committed to the idea that our mathematical theories describe unique collections of mathematical objects. The commitment is to the existence of *all* the mathematical objects that could possibly exist. Thus the view is committed to all the mathematical objects in all the models of the natural numbers, even non-intended models. Similarly, full-blooded Platonism is committed to the existence of ZFC sets, and various extensions of ZFC, as well as the objects in various alternative set theories. The question of which are the real sets is easily answered: they are all real. We might make exceptions for the more familiar mathematical theories such as arithmetic, where we might insist that such theories are about only the intended models. Again, there are details to be tidied up, but you get the basic idea.[3]

3.1.2 Structuralism

Structuralism in the philosophy of mathematics is the view that the proper subject matter of mathematics is the relationships between various kinds of mathematical entities, rather than the entities themselves. So, for instance, number theory is the study of ω-sequences, not the study of any particular ω-sequence. Moreover, the nature of the entities that constitute the ω-sequences is irrelevant – what is important is the structure that is common

---

[3] Other plenitudinous positions in the philosophy of mathematics have been advanced, for example, by Zalta (1983) and Priest (2005). Some of the theories in question (at least on some readings) are Platonist, but others pursue the Meinongian idea that we can refer to non-existent objects. The latter are, in one sense, plenitudinous, but it's not right to think of them as Platonist.

to all such sequences. There is something undeniably right about structuralism. Indeed, the structuralist slogans, 'mathematical objects are places
in structures' or, Michael Resnik's (1997) favourite, 'mathematics is the
science of patterns' seem to reveal important insights into the nature of
mathematics and its subject matter.

The real significance of structuralism, though, is in its ability to provide
answers to some rather difficult problems in the philosophy of mathematics. For example, structuralism is able to explain why mathematicians are
typically only interested in describing the objects they study up to isomorphism – for that is all there is to describe. Structuralism is also able to
explain why we're inclined to think that either all of the natural numbers
exist or none of them exist – the natural numbers, like other structures,
come as a package. Most importantly, structuralism provides a neat and
plausible response to Benacerraf's underdetermination problem. According to structuralists, anything can play the role of the number 2, so long
as the object in question has the appropriate relationships with other enti-
ties, and together they make up an ω-sequence. Von Neumann's $\{\emptyset, \{\emptyset\}\}$,
Zermelo's $\{\{\emptyset\}\}$, or even the Roman emperor Julius Caesar can play the
role of the number 2 (so long as Caesar has the appropriate structural
relations with enough other items of an ω-sequence). According to the
structuralist, however, neither Julius Caesar nor $\{\emptyset, \{\emptyset\}\}$ nor $\{\{\emptyset\}\}$ *is* the
number 2, for the number 2 is no more and no less than a position in a
structure.[4]

Think of positions in a sporting team. Anyone can play these positions.
Indeed, asking a question such as 'what colour hair does a baseball shortstop
have?' are misguided. A baseball shortstop is nothing more than someone
in a baseball fielding line-up who stands between the second and third base.
The details of the individual are not part of the game. Just as in the mathematics case, where a variety of entities can play the role of 2, anyone can
play the shortstop role. And also, like the mathematical case, no particular
person *is* the shortstop.

---

4 Structuralism has had its fair share of supporters. These have included mathematicians
  of the calibre of Dedekind, Hilbert, and Poincaré, and philosophers such as Quine
  and Benacerraf and (somewhat tentatively) Putnam. The most detailed philosophical
  treatment of the position is to be found in the work of Michael Resnik (1997) and
  Stewart Shapiro (1997).

But as you might expect, there are factions here as well. We can ask whether the structures in question need to be instantiated or not. A Platonic position would not require there to be instantiations of the structure in question. According to this view, it doesn't matter whether there are physical instantiations of ω-sequences. The natural number structure is not to be held hostage to the way the world happens to be. Shapiro calls this position *ante rem* structuralism. A more Aristotelian position (which Shapiro calls *in re* structuralism) holds that only instantiated structures exist. There are costs and benefits to be considered here. The *in re* position is more modest; it does not posit anything other than the kinds of structures that are in fact found in the world. It thus has the virtue of simplicity and looks as though it is well placed to provide a workable epistemology for mathematics. The problem, though, is that the world may not provide rich enough structures for the mathematics we know and love. For example, think what it would mean for the real-number structure to be instantiated. For the instantiation to be in space, we'd require space to be continuous and unbounded. It is far from clear that this is the case. There may in fact be no instantiations of the real numbers. And there may even be problems finding enough objects for an instantiation of the natural numbers. So without further work, *in re* structuralism is too modest – it may not give us enough structures for even the most basic mathematics. The alternative, *ante rem* structuralism, has no such problems. It posits any structure instantiated or not. Every consistent structure is real. So there are no modesty problems here. But, then, the familiar epistemic problem is: how do we know about such uninstantiated structures?[5]

3.2 Indispensability arguments

One of the most intriguing features of mathematics is its applicability to empirical science. Every branch of science draws upon large and often diverse portions of mathematics, from the use of Hilbert spaces in quantum mechanics to the use of differential geometry in general relativity. It's

---

[5] There are yet other versions of structuralism, according to which the structures in which mathematicians are interested are possible structures (Hellman 1989). Whether such positions count as realist or not depends on your position on possibilia. There are both realist and anti-realist positions to be defended here.

not just the physical sciences that avail themselves of the services of mathematics, either. Ecology, for instance, makes extensive use of differential equations and statistics. The roles mathematics plays in these theories are also varied. Not only does mathematics help with empirical predictions, it allows the elegant and economical statement of many theories. Indeed, so important is the language of mathematics to science, that it is hard to imagine how theories such as quantum mechanics and general relativity could even be stated without employing a substantial amount of mathematics.

From the rather remarkable but seemingly uncontroversial fact that mathematics is indispensable to science, an argument with a serious metaphysical conclusion can be made. In particular, W. V. Quine (1953, 1976, 1981a, 1981b) and Hilary Putnam (1971, 1979) have argued that the indispensability of mathematics to empirical science gives us good reason to believe in the existence of mathematical entities. According to this line of argument, reference to (or quantification over) mathematical entities such as sets, numbers, functions, and such is indispensable to our best scientific theories, and so we ought to be committed to the existence of these mathematical entities. To do otherwise is to be guilty of what Putnam has called 'intellectual dishonesty' (Putnam 1971, p. 347). Moreover, mathematical entities are seen to be on an epistemic par with the other theoretical entities of science, since belief in the existence of the former is justified by the same evidence that confirms the theory as a whole (and hence belief in the latter). This argument is known as the Quine–Putnam indispensability argument for mathematical realism. There are other indispensability arguments,[6] but this one is by far the most influential, and so in what follows I'll concentrate on it.

[6] In general, an indispensability argument is an argument that purports to establish the truth of some claim based on the indispensability of the claim in question for certain purposes (to be specified by the particular argument). For example, if explanation is specified as the purpose, then we have an explanatory indispensability argument. Thus we see that inference to the best explanation is a special case of an indispensability argument. See the introduction to Field (1989, pp. 14–20) for a nice discussion of indispensability arguments and inference to the best explanation. See also Maddy (1992) and Resnik (1995) for variations on the Quine–Putnam version of the argument. I should add that although the version of the argument presented here is generally attributed to Quine and Putnam, it differs in a number of ways from the arguments advanced by either Quine or Putnam. Still, the version presented here is the one generally discussed in the contemporary literature.

The indispensability argument has attracted a great deal of attention, in part because many see it as the best argument for mathematical realism (or Platonism). Thus anti-realists about mathematical entities (or nominalists) need to identify where the indispensability argument goes wrong. Many Platonists, on the other hand, rely on this argument to justify their belief in mathematical entities. The argument places nominalists who wish to be realist about other theoretical entities of science (quarks, electrons, black holes and such) in a particularly difficult position. For typically they accept something quite like the indispensability argument as justification for realism about quarks and black holes. Most scientific realists accept inference to the best explanation. Indeed, inference to the best explanation is arguably the cornerstone of scientific realism. But inference to the best explanation may be seen as a kind of indispensability argument, so any realist who accepts the former while rejecting the latter finds themself in a very unstable position. They would seem to be holding a double standard about ontology.

The indispensability argument can be stated in the following explicit form:

(P1) We ought to have ontological commitment to all and only the entities that are indispensable to our current best scientific theories.
(P2) Mathematical entities are indispensable to our best scientific theories.
(C) We ought to have ontological commitment to mathematical entities.

Thus formulated, the argument is valid. This forces the focus onto the two premises. In particular, a couple of important questions naturally arise. The first concerns how we are to understand the claim that mathematics is indispensable. Another question concerns the first premise. This premise has a great deal packed into it and clearly requires some defence. We'll consider each of these questions in turn before turning to objections to the argument.

3.2.1 What is it to be indispensable?

The question of how we should understand 'indispensability' is crucial. Quine spells this out in terms of the entities quantified over in the canonical first-order formulation of our best scientific theories. But we don't have to follow him on this. After all, it is not clear that all scientific theories can be put into this form. In any case, it is useful to try to spell out the notion of 'indispensability' in more intuitive terms.

The first thing to note is that 'dispensability' is not the same as 'eliminability', otherwise every entity would be dispensable (due to Craig's Theorem).[7] What we require for an entity to be 'dispensable' is for it to be eliminable and that the theory resulting from the entity's elimination be an attractive theory. (Perhaps, even stronger, we require that the resulting theory be more attractive than the original.) We will need to spell out what counts as an attractive theory, but for this we can appeal to the standard desiderata for good scientific theories: empirical success; unificatory power; simplicity; explanatory power; fertility; and so on. Of course there will be debate over what desiderata are appropriate and over their relative weightings, but such issues need to be addressed and resolved independently of issues of indispensability (Burgess 1983).

The question of how much mathematics is indispensable (and hence how much mathematics carries ontological commitment) naturally arises at this point. It seems that the indispensability argument only justifies belief in enough mathematics to serve the needs of science. Thus we find Putnam speaking of 'the set theoretic "needs" of physics' (Putnam 1971, p. 346) and Quine claiming that the higher reaches of set theory are 'mathematical recreation ... without ontological rights' (Quine 1986, p. 400), since these parts of set theory do not find physical applications. One could take a less restrictive line and claim that the higher reaches of set theory, although without physical applications, do carry ontological commitment by virtue of the fact that they have applications in other parts of mathematics. So long as the chain of applications eventually bottoms out in physical science, we could rightly claim that the whole chain carries ontological commitment. Quine himself justifies some transfinite set theory along these lines, but he sees no reason to go beyond the constructible sets (Quine 1986, p. 400).

---

[7] This theorem states that relative to a partition of the vocabulary of an axiomatisable theory T into two classes, t and o (theoretical and observational, say), there exists an axiomatisable theory T^* in the language whose only non-logical vocabulary is o, of all and only the consequences of T that are expressible in o alone. If the vocabulary of the theory can be partitioned in the way that Craig's Theorem requires, then the theory can be reaxiomatised so that apparent reference to any given theoretical entity is eliminated. See Field (1980, p. 8) for further details and for its relevance to the indispensability argument.

His reasons for this restriction, however, have little to do with the indispensability argument, so again we do not need to follow Quine on this issue.

3.2.2 Naturalism and holism

Although both premises of the indispensability argument have been questioned, it's the first premise that is most obviously in need of support. This support comes from the doctrines of naturalism and holism.

Following Quine, naturalism is usually taken to be the philosophical doctrine that there is no first philosophy and that the philosophical enterprise is continuous with the scientific enterprise (Quine 1981b). By this Quine means that philosophy is neither prior to nor privileged over science. What is more, science, thus construed, is taken to be the complete story of the world. Naturalism arises out of a deep respect for scientific methodology and an acknowledgment of the undeniable success of this methodology as a way of answering fundamental questions about all manner of things. As Quine suggests, its source lies in 'unregenerate realism, the robust state of mind of the natural scientist who has never felt any qualms beyond the negotiable uncertainties internal to science' (Quine 1981a, p. 72). For the metaphysician this means looking to our best scientific theories to determine what exists, or, perhaps more accurately, to determine what we ought to believe to exist. In short, naturalism rules out unscientific ways of determining what exists. For example, naturalism rules out believing in angels for mystical reasons. Naturalism would not, however, rule out belief in angels if our best scientific theories were to require them.

Naturalism, then, gives us a reason for believing in the entities in our current best scientific theories and no other entities. Depending on exactly how you conceive of naturalism, it may or may not tell you whether to believe in all the entities of your best scientific theories. I take it that naturalism does give us some reason to believe in all such entities, but that this is defeasible. This is where holism comes in.

Confirmational holism is the view that theories are confirmed or disconfirmed as wholes (Quine 1953, p. 41). So, if a theory is confirmed by empirical findings, the whole theory is confirmed. In particular, whatever mathematics is made use of in the theory is also confirmed (Quine 1976, pp. 120–2). Furthermore, it is the same evidence that is appealed to in justifying belief in the mathematical components of the theory that is appealed to in justifying the empirical portion of the theory (if indeed the empirical can be separated from the mathematical at all). Naturalism and holism taken together then justify (P1). Roughly, naturalism gives us the 'only' and holism gives us the 'all' in (P1).

It is worth noting that in Quine's writings there are at least two holist themes. The first is the confirmational holism discussed above (often called the Quine–Duhem thesis).[8] The other is semantic holism, which is the view that the unit of meaning is not the single sentence, but systems of sentences (and in some extreme cases the whole of language). This latter holism is closely related to Quine's well-known denial of the analytic–synthetic distinction (Quine 1953) and his thesis of the indeterminacy of translation (Quine 1960). Although for Quine, confirmational holism and semantic holism are closely related, there is good reason to distinguish them, since the latter is generally thought to be highly controversial, while the former is arguably less so. This is important to the present debate because Quine explicitly invokes the controversial semantic holism in support of the indispensability argument (Quine 1953, pp. 45–6). Most commentators, however, are of the view that only confirmational holism is required to make the indispensability argument fly (see, for example, Field (1989, pp. 14–20); Maddy (1992); and Resnik (1995; 1997)). It should be kept in mind, however, that while the argument, thus construed, is Quinean in flavour, it is not, strictly speaking, Quine's argument.

3.3 Objections

There have been many objections to the indispensability argument, including Charles Parsons's (1980) complaint that the obviousness of basic mathematical statements is left unaccounted for by the Quinean picture and Philip Kitcher's (1984, pp. 104–5) complaint that the indispensability argument doesn't explain why mathematics is indispensable to science. The objections that have received the most attention, however, are those due to Hartry Field, Penelope Maddy, and Elliott Sober. In particular, over the last

---

[8] See Duhem (1954) – originally published in 1906 – for an early, classic presentation of confirmational holism.

20–30 years, Field's nominalisation programme has dominated discussions of the ontology of mathematics.

We will consider Field's (1980) positive proposal in more detail in the next chapter, but for now let's consider his objection to the indispensability argument. In a nutshell, Field denies the second premise of the Quine–Putnam argument. That is, he suggests that despite appearances, mathematics is not indispensable to science. But of course he cannot simply deny this and leave it there; he must make a case for the dispensability of mathematics. There are thus two parts to Field's project. The first is to attempt to demonstrate that our best scientific theories can, indeed, survive without mathematics.[9] That is, he attempts to show that we can do without quantification over mathematical entities and that what we are left with are reasonably attractive theories. To this end he attempts to nominalise a large fragment of Newtonian gravitational theory. Although this is a far cry from showing that all our current best scientific theories can be nominalised, it is certainly not trivial. The hope is that once one sees how the elimination of reference to mathematical entities can be achieved for a typical physical theory, it will seem plausible that this could be extended to the rest of science.

The second part of Field's project is to argue that mathematical theories don't need to be true to be useful in applications – they need merely be conservative. (This is, roughly, that if a mathematical theory is added to a nominalist scientific theory, no nominalist consequences follow that wouldn't follow from the nominalist scientific theory alone.) If, indeed, mathematics is conservative, that would explain why mathematics can be used in science, but it would not explain why it is used. The latter is due to the fact that mathematics makes calculation and statement of various theories much simpler. Thus, for Field, the utility of mathematics is merely pragmatic – mathematics is not indispensable after all.

There has been a great deal of debate over the prospect for the success of Field's programme but few have doubted its significance. As I said, we will return to Field's proposal in the next chapter. For now, we simply note that Field's programme, among other things, amounts to a very interesting line of objection to the indispensability argument. He gives us reason

---

[9] This is known as attempting to *nominalise* the science in question, since the exercise is to rid science of the nominalistically dubious mathematical entities.

for pause on what was once thought to be the uncontroversial part of the indispensability argument: the indispensability thesis itself.

Penelope Maddy takes a different line of attack. She points out that if (P1) is false, Field's project may turn out to be irrelevant to the realism–anti-realism debate in mathematics. To this end, Maddy presents a number of serious objections to the first premise of the indispensability argument (Maddy 1992; 1995; 1997). In particular, she suggests that we ought not have ontological commitment to all the entities indispensable to our best scientific theories. Her objections draw attention to problems of reconciling naturalism with confirmational holism. In particular, she points out how a holistic view of scientific theories has problems explaining the legitimacy of certain aspects of scientific and mathematical practices – practices which, presumably, ought to be legitimate given the high regard for scientific practice that naturalism recommends. It is important to appreciate that her objections, for the most part, are concerned with methodological consequences of accepting the Quinean doctrines of naturalism and holism – the doctrines used to support the first premise. The first premise is thus called into question by undermining its support.

It is also worth commenting on the force of the argument style Maddy adopts. She argues for an internal inconsistency in the Quinean background. She embraces the Quinean worldview, for the sake of argument, and shows that this leads to trouble. This is very different from, and much more compelling than, for example, arguing that from the point of view of some other metaphysical position, the Quinean worldview is wrong. Maddy attempts to show that the Quineans, *by their own lights*, cannot accept both naturalism and holism. I'm not suggesting that she succeeds in undermining the Quinean position – I think there are responses the Quinean can make and ultimately I stand on the side of Quine in this debate. I am just commenting on the argumentative strategy Maddy adopts and drawing attention to the strength of such a strategy. She is not begging questions or merely attempting to push the burden of argumentative proof around. She takes Quine on, on his own terms.

Maddy's first objection to the indispensability argument is that the attitudes of working scientists towards the components of well-confirmed theories vary from belief, through tolerance, to outright rejection (Maddy

1992, p. 280). She illustrates this with a case study from the history of science. The example concerns the debate over (modern) atomic theory from the early nineteenth century to the early twentieth century. From about 1860 the atom became the fundamental unit of chemistry, but the existence of atoms was not universally accepted until early in the twentieth century, when Albert Einstein's (1879–1955) theoretical work and Jean Baptiste Perrin's (1870–1942) experimental work on Brownian motion sealed the deal and it was no longer possible to deny the reality of atoms. But the Quinean picture of science would, it seems, have it that atoms ought to have been accepted as real from about 1860, despite renowned scientists such as Poincaré and Wilhelm Ostwald (1853–1932) remaining sceptical of the reality of atoms until as late as 1904. The point is that naturalism counsels us to respect the methods of working scientists, and yet holism is apparently telling us that working scientists are wrong in allowing the variety of attitudes they have towards the entities in their theories. Maddy suggests that we should side with naturalism not holism here. Thus we should endorse the attitudes of working scientists who apparently do not believe in all the entities posited by our best theories. We should thus reject (P1) of the indispensability argument.

The next objection follows from the first. Once one rejects the picture of scientific theories as homogeneous units, the question arises whether the mathematical portions of theories fall within the true elements of the confirmed theories or within the idealised elements. Maddy suggests the latter. Her reason for this is that scientists themselves do not seem to take the indispensable application of a mathematical theory to be an indication of the truth of the mathematics in question. For example, the false assumption that water is infinitely deep is often invoked in the analysis of water waves, or the assumption that matter is continuous is commonly made in fluid dynamics (Maddy 1992, pp. 281–2). Such examples indicate that scientists will invoke whatever mathematics is required to get the job done, without regard to the truth of the mathematical theory in question (Maddy 1995, p. 255). Again, it seems that confirmational holism is in conflict with actual scientific practice, and hence with naturalism. And again Maddy sides with naturalism. The point here is that if naturalism counsels us to side with the attitudes of working scientists in such matters, then it seems that we ought not take the indispensability of some mathematical theory in a physical application as an indication of the truth of the mathematical theory. Furthermore, since we have no reason to believe that the mathematical theory in question is true, we have no reason to believe that the entities posited by the (mathematical) theory are real. So once again we ought to reject (P1).

Maddy's third objection is that, from a Quinean point of view, it is hard to make sense of what working mathematicians are doing when they try to settle independent questions. Recall that these are questions that are independent of the standard axioms of set theory – the ZFC axioms. As we saw in Chapter 2, the question of the size of the continuum – whether the continuum hypothesis is true – is one such question. In order to settle some of these independent questions, new axiom candidates have been proposed to supplement ZFC, and arguments have been advanced in support of these candidates. The problem is that the arguments advanced seem to have nothing to do with applications in physical science: they are typically intra-mathematical arguments. According to the thinking lying behind the indispensability argument, the new axioms should be assessed on how well they cohere with our current best scientific theories. That is, set theorists should be assessing the new axiom candidates with one eye on the latest developments in physics. Given that set theorists do not do this, confirmational holism seems to be advocating a revision of standard mathematical practice, and this too, claims Maddy, is at odds with naturalism (Maddy 1992, pp. 286–9).

Although Maddy does not formulate this last objection in a way that directly conflicts with (P1), it certainly illustrates a tension between naturalism and confirmational holism. And since both these are required to support (P1), the objection indirectly casts doubt on (P1). Maddy, however, endorses naturalism and so takes the objection to demonstrate that confirmational holism is false.

Elliott Sober has similar concerns about the indispensability argument and arrives at a similar conclusion to Maddy. Sober (1993) takes issue with the claim that mathematical theories share the empirical support accrued by our best scientific theories. In essence, he argues that mathematical theories are not being tested in the same way as the clearly empirical theories of science. He points out that hypotheses are confirmed relative to competing hypotheses. Thus if mathematics is confirmed along with our best empirical hypotheses (as the indispensability argument has it), there must be mathematics-free competitors. But Sober points out that all scientific theories employ a common mathematical core. Thus, since there are no competing hypotheses, it is a mistake to think that mathematics receives confirmational support from empirical evidence in the way other scientific hypotheses do.

This in itself does not constitute an objection to (P1) of the indispensability argument, as Sober is quick to point out (Sober 1993, p. 53), although it does constitute an objection to Quine's overall view that mathematics is part of empirical science. As with Maddy's third objection, it gives us a reason to reject confirmational holism. The impact of these objections on (P1) depends on how crucial you think confirmational holism is to that premise. Certainly much of the intuitive appeal of (P1) is eroded if confirmational holism is rejected. In any case, to subscribe to the conclusion of the indispensability argument in the face of Sober's or Maddy's objections is to hold the position that it's permissible to have ontological commitment to entities that receive no empirical support. This, if not outright untenable, is certainly not in the spirit of the original Quine–Putnam argument.

It is not clear how damaging the above criticisms are to the indispensability argument. Indeed, the debate is very much alive, with many recent articles devoted to the topic. Closely related to this debate is the question of whether there are any other cogent arguments for Platonism. If, as some believe, the indispensability argument is the only argument for Platonism worthy of consideration, then if it fails, Platonism in the philosophy of mathematics is bankrupt. Of relevance, then, is the status of other arguments for and against mathematical realism. In any case, it is worth noting that the indispensability argument is one of a small number of arguments that have dominated discussions of the ontology of mathematics. It is therefore important that this argument not be viewed in isolation. It is also important to note that even if the indispensability argument is the only good argument for Platonism, the failure of this argument does not necessarily authorise nominalism, for the latter, too, may be without support. It does seem fair to say, however, that if the objections to the indispensability argument are sustained then one of the most important arguments for Platonism is undermined. This would leave Platonism on rather shaky ground. We look at some of the nominalist strategies in the philosophy of mathematics in the next chapter.

Discussion questions

1. According to structuralists, the number 2 is just a position in a structure. But which structure? The natural number 2 is part of the natural number structure, but it is also part of the structure that is the integers. Moreover, the real number 2 is part of the real number structure, and the complex number $2 + 0i$ is part of the complex number structure. Does this mean that the structuralist is committed to a plurality of 2s? How can they make sense of the idea that the natural numbers are a subset of the real numbers, if each number depends on its structure for its identity?

2. Consider how a full-blooded Platonist might respond to an objection that with so many mathematical objects around, we can no longer secure reference to the ones we're interested in. Recall that on the assumption that all Nepalese villages exist, my beliefs about a particular Nepalese village can't fail to be true of one of the villages. But which one? What right do I have to say '*the* Nepalese village I'm thinking of has a population of 370'? Do such plenitudinous theories solve the epistemological access problem, but at the expense of a reference problem?

3. Really full-blooded Platonists posit the existence of mathematical entities that inhabit inconsistent theories as well as those that inhabit the consistent theories. So, they accept inconsistent mathematical entities, such as the Russell set, along with the usual sets of ZFC. Can you think of any motivation to take this extra step? Can the full-blooded Platonist tell you which theories are consistent?

4. How might an anti-realist about science resist the conclusion of the indispensability argument?

5. How might a defender of the indispensability argument reply to Maddy's objections? Is it really a violation of naturalism to criticise scientists – even scientists of the calibre of Poincaré and Ostwald?

6. Is Sober right that scientific hypotheses are only ever confirmed relative to competing hypotheses? What if there is only one hypothesis and it exactly predicts all experimental results?

7. Think about the kind of mathematical realism delivered by the indispensability argument. Can the argument be used by any of the versions of mathematical realism we've seen so far? Can the indispensability

argument be used to support a mathematical realism that holds that mathematical truths are necessary truths and that mathematical entities exist of necessity?

Recommended further reading

For more on plenitudinous views of mathematics see Balaguer (1998, Chapters 3 and 4), and Priest (2005). For more on structuralism see Resnik (1997) and Shapiro (1997).

Although the indispensability argument is to be found in many places in Quine's writings (including 1953, 1976), the *locus classicus* is Putnam's short monograph *Philosophy of Logic* (included as a chapter of the second edition of the first volume of his collected papers (Putnam 1971)). See also the introduction of Field (1989), which has an excellent outline of the argument along with a good, accessible overview of his strategy for attacking the indispensability argument. Colyvan (2001) is a sustained defence of the argument and includes replies to the objections discussed in this chapter. For an interesting pragmatic variant of the indispensability argument see Resnik (1995). This version allegedly does not require the assumption of confirmational holism.

The Maddy and Sober criticisms of the argument are found in Maddy (1992) and Sober (1993).

Balaguer, M. 1998. *Platonism and Anti-Platonism in the Philosophy of Mathematics*, New York: Oxford University Press.

Colyvan, M. 2001. *The Indispensability of Mathematics*, New York: Oxford University Press.

Field, H. 1989. *Realism, Mathematics and Modality*, Oxford: Blackwell.

Maddy, P. 1992. 'Indispensability and Practice', *Journal of Philosophy*, 89(6): 275–89.

Priest, G. 2005. *Towards Non-Being: The Logic and Metaphysics of Intentionality*, Oxford: Clarendon Press.

Putnam, H. 1971. *Philosophy of Logic*, New York: Harper, reprinted in *Mathematics, Matter and Method: Philosophical Papers, vol. I*, 2nd edn, Cambridge University Press, 1979, pp. 323–57.

Quine, W. V. 1953. 'Two Dogmas of Empiricism', in *From a Logical Point of View*, Cambridge, MA: Harvard University Press, pp. 20–46.

1976. 'Carnap and Logical Truth', reprinted in *The Ways of Paradox and Other Essays*, revised edn, Cambridge, MA: Harvard University Press, pp. 107–32 (and in Benacerraf and Putnam (eds.), pp. 355–76).

Resnik, M. D. 1995. 'Scientific vs. Mathematical Realism: The Indispensability Argument', *Philosophia Mathematica*, 3(2): 166–74.

1997. *Mathematics as a Science of Patterns*, Oxford: Clarendon Press.

Shapiro, S. 1997. *Philosophy of Mathematics: Structure and Ontology*, New York: Oxford University Press.

Sober, E. 1993. 'Mathematics and Indispensability', *Philosophical Review*, 102(1): 35–57.

4 Fiction, metaphor, and partial truths

Mathematics may be defined as the subject in which we never know what we are talking about, nor whether what we are saying is true.

Bertrand Russell (1872–1970)[1]

In the last chapter we saw one of the main cases for Platonism, namely, the indispensability argument. In this chapter we look at a few anti-realist philosophies of mathematics. Each of these positions can be understood as a response to the indispensability argument. They are also motivated by the Benacerraf epistemic challenge to Platonism and the hope that it's easier to be rid of troublesome mathematical entities than it is to provide a Platonist epistemology.

4.1 Fictionalism

Fictionalism in the philosophy of mathematics is the view that mathematical statements, such as '$7+5 = 12$' and 'π is irrational', are to be interpreted at face value and, thus interpreted, are false. Fictionalists are typically driven to reject the truth of such mathematical statements because these statements imply the existence of mathematical entities, and, according to fictionalists, there are no such entities. Fictionalism is a nominalist (or anti-realist) account of mathematics in that it denies the existence of a realm of abstract mathematical entities. It should be contrasted with mathematical realism (or Platonism), where mathematical statements are taken to be true and, moreover, are taken to be truths about mathematical entities. Fictionalism should also be contrasted with other nominalist philosophical

---

[1] B. Russell, 'Recent Work on the Principles of Mathematics', *International Monthly*, 4 (1901): 83–101. Reprinted as 'Mathematics and the Metaphysicians', in B. Russell, *Mysticism and Logic*, London: Longmans Green, 1918, pp. 74–86.

accounts of mathematics that propose a reinterpretation of mathematical statements, according to which the statements in question are true but no longer about mathematical entities. Fictionalism is thus an error theory of mathematical discourse: at face value mathematical discourse commits us to mathematical entities, and although we normally take many of the statements of this discourse to be true, in doing so we are in error.

Although fictionalism holds that mathematical statements implying the existence of mathematical entities are strictly speaking false, there is a sense in which these statements are true – they are true in the story of mathematics. The idea here is borrowed from literary fiction, where statements such as 'Bilbo Baggins is a hobbit' are strictly speaking false (because there are no hobbits). But such statements are true in Tolkien's fiction *The Hobbit*. Fictionalism about mathematics shares the virtue of ontological parsimony with other nominalist accounts of mathematics. It also lends itself to a very straightforward epistemology: there is nothing to know beyond the human-authored story of mathematics. And coming to know the various fictional claims requires nothing more than knowledge of the story in question. The most serious problem fictionalism faces is accounting for the applicability of mathematics. Mathematics, unlike Tolkien's stories, is apparently indispensable to our best scientific theories and, as we have seen, this suggests that we ought to be realists about mathematical entities.

4.1.1 Motivation for fictionalism

As we have seen there are two competing pressures in finding an adequate philosophy of mathematics (Benacerraf 1983b). The first is to provide a uniform semantics across mathematical discourse and non-mathematical discourse. We want sentences such as '4 is larger than 2' to be treated semantically in the same way as sentences such as 'Sydney is larger than Amsterdam', for at face value these two sentences seem to have the same structure and ought to have similar truth conditions. The second pressure is to provide an adequate naturalistic epistemology for mathematics, one that does not make a mystery of how we come by mathematical knowledge. And recall that these are usually taken to be competing pressures because realist philosophies of mathematics have little problem providing a uniform semantics but typically have trouble providing a naturalistically acceptable epistemology. Nominalist philosophies of mathematics, on the other hand, typically have difficulty providing a uniform semantics, with many nominalist philosophies having to give up on this entirely. But nominalist accounts fare much better with epistemology, for according to these theories, mathematical knowledge – whatever it is – is not knowledge of abstract entities.

Fictionalist philosophies of mathematics can be seen to be providing an elegant way of dealing with these two competing pressures. Fictionalism does employ a uniform semantics. '4 is larger than 2' is read at face value in the obvious way, just as 'Sydney is larger than Amsterdam' is. The difference, according to fictionalism, is that the latter sentence is true but the former is false. '4 is larger than 2' is taken to be false because there are no referents for '4' and '2'. But the semantics in both cases is the same. As with other nominalist theories of mathematics, epistemology does not present any serious difficulties for fictionalism. According to fictionalism, there is no mathematical knowledge apart from knowledge of the fiction of mathematics itself. Knowing that in the story of mathematics $7+5 = 12$ is no more problematic than knowing that in the Tolkien story Bilbo Baggins is a hobbit. In both cases we know this by reading the relevant stories, listening to others who are well versed in the stories in question or, more adventurously, by exploring the logical consequences of the respective stories.

The price of fictionalism, however, is that much of mathematics is taken to be false.[2] While at first it might seem unintuitive to claim that '$7+5 = 12$' is false, to claim otherwise is to commit oneself to the existence of numbers. After all, it follows straightforwardly from '$7 + 5 = 12$' that, for example, there exist numbers x and y, such that $x+y = 12$. According to fictionalists, the existence of mathematical objects is problematic enough to warrant denying the truth of such statements.

It is also important to note that fictionalism in mathematics does not mean that anything goes. Authors of mathematical theories, like writers of good literary fiction, are not free to develop their fiction in any way they please. For a start, consistency is usually thought to be strongly desirable.

---

[2] I say 'much of mathematics' because fictionalism does preserve the truth values of negative existentials like 'there is no largest prime number'. This statement is true in standard mathematics, and therefore true in the story of mathematics, but according to fictionalism it is also true simpliciter, because there are no numbers and a fortiori there is no largest prime.

Beyond that, there are also requirements not to introduce unnecessary items. In good mathematics, as in good literary fiction, posited entities contribute to the story. But perhaps the greatest constraint on writing mathematical fiction is that the latest instalment must square with all previous instalments. Previous generations of mathematicians introduced such 'characters' as sets, functions, natural numbers, and so on. The current generation of mathematicians must develop these 'characters' in ways that are consistent with what went before. It is as though current mathematicians are all contributing to a multi-authored series of books. Just as J. R. R. Tolkien was heavily constrained in the last book in *The Lord of the Rings* trilogy by what went before in *The Hobbit* and the previous two *The Lord of the Rings* books, so too modern mathematicians cannot develop the fiction of mathematics in any way they please.

4.1.2 The challenge for fictionalism

The biggest problem facing fictionalism in mathematics is to explain the central role mathematics plays in scientific inquiry. As we have seen in the previous chapter, there is a very influential argument for the existence of mathematical entities that needs to be confronted: the indispensability argument.

Recall that this argument (outlined on p. 43) is supposed to resemble a style of argument endorsed by scientific realists. According to scientific realists, we are committed to electrons, black holes, and other unobservable theoretical entities because of the role those entities play in our best scientific theories. The indispensability argument can be thought of as an attempt to push scientific realists a bit further – to mathematical realism – but seen in this light it is likely to have little bite on scientific anti-realists. As we have already seen, there is an assumption of a naturalistic attitude, whereby we are encouraged to look towards science for answers to questions about ontology. This much is typically endorsed by scientific realists. But Premise 1 of the argument says more: it also suggests that we ought to believe in *all* the entities of our best scientific theories. This gives voice to a kind of holism about scientific theories, whereby we cannot pick and choose among the parts of our best scientific theories. According to the holism in question, we believe our best confirmed theories in their entirety. The kind of holism involved is confirmational holism, which has it that theories are confirmed or disconfirmed as wholes, not one hypothesis at a time.

It is clear that fictionalists have two basic options: deny Premise 1 or deny Premise 2. The first option typically involves giving up holism, although usually this is attempted while trying to maintain a commitment to naturalism and scientific realism. The second option involves showing that, in the relevant respect, mathematics is dispensable to science. This brings us back to Hartry Field's (1980) heroic attempt to do science without numbers.

4.1.3 The hard road to nominalism: Hartry Field's fictionalism

Field's project is motivated by a commitment to providing a uniform semantics and by epistemological concerns with Platonism. The particular approach Field adopts is also motivated by a couple of other considerations: a commitment to providing intrinsic explanations (i.e., explanations that do not rely on extraneous entities) and the elimination of arbitrariness from scientific theories (e.g., the elimination of conventional coordinate frames and units of distance). While these considerations provide the motivation for fictionalism, according to Field, an adequate fictionalist philosophy of mathematics must explicitly address the indispensability argument. For the latter he takes to be the only good argument for Platonism, and, as such, it presents a serious obstacle to any nominalist philosophy of mathematics: undermine the indispensability argument and you undermine Platonism. It is Field's willingness to take the indispensability argument head-on that gives his account its distinctive flavour. And it is for this reason that I refer to it as the 'hard road' to nominalism. Just how hard this road is will become apparent as we get into the details of the Field programme.

According to all varieties of mathematical fictionalism, most of accepted mathematics is strictly speaking false, but true in the fictional story of mathematics. Field, however, recognises that the fictionalist account cannot stop there. After all, why should this particular fiction – the fiction of standard mathematics – prove to be in such demand in science? Field's answer to this question is ingenious. He simultaneously suggests how mathematics might be dispensed with and how, despite its dispensability, it could be used so fruitfully in its various applications throughout science.

The first part of Field's project – showing the dispensability of mathematics – begins by showing how a typical scientific theory such as Newtonian gravitational theory might be constructed without quantifying over mathematical items. The basic idea is to be a substantivalist about space-time points[3] and then work directly with space-time points/regions. Instead of talking of the gravitational potential, for example, of some space-time point, Field compares space-time points with respect to their gravitational potential. The former, standard way of talking (in terms of gravitational potential of space-time points) involves a gravitational potential function which is a map from the space-time manifold to real numbers and this seems to commit one to realism about space, time, functions, and the real numbers. But Field, following a lead from David Hilbert, notices that one can do all one wants merely by comparing space-time points with respect to their gravitational potential. This relational approach does away with the nominalistically unacceptable mathematical machinery (functions and real numbers) in the theory itself. But Field also proves a representation theorem that shows that in the metatheory one can recover all the relevant numerical claims. In particular, in the space-time theory Field considers (a fragment of Newtonian gravitational theory), there are no gravitational potential functions, mass-density functions or spatio-temporal coordinate functions, but the representation theorem guarantees that these are recoverable in the metatheory. So, in a very important sense, nothing is lost.

Field, however, does not advocate doing science without mathematics; it is just that science *can* be done without mathematics. And the latter is enough to suggest that mathematics is dispensable to science. But now the question arises as to why invoking the fiction of mathematics does not lead to trouble. After all, combining a scientific theory with a work of fiction would generally lead to all sorts of false and perhaps even contradictory results. What is so special about mathematics and why is it acceptable to continue using the fiction of mathematics? This brings us to the second part of Field's programme: demonstrating the conservativeness of mathematics. Recall that the idea here is to show that a mathematical theory, when

---

[3] Substantivalism about space-time points is committed to the reality of the space-time manifold and, in particular, to the reality of space-time points. This view should be contrasted with relational accounts of space-time, which have it that space-time is just a way of organising events spatially and temporally – there is no reason to posit space-time in addition to the events themselves.

combined with any nominalistic scientific theory, does not yield nominalistic consequences that could not have been derived from the nominalsitic theory alone. The mathematics allows for easier derivations and the like, but enlisting it in the services of science does not yield anything new about the physical world. Put figuratively, the falsity of the mathematics does not infect the science that employs it. So if mathematics is conservative, we can continue using it and no damage will be done. The conservativeness claim is thus crucial in maintaining Field's contention that his fictionalism does not result in any change to scientific practice.

Why believe that mathematics is conservative? Field provides a couple of different proofs of this, but the intuitive argument in support of the claim will suffice here. Field argues that good mathematics is conservative, and a discovery that a mathematical theory was not conservative would be a reason to revise the mathematical theory in question. After all, if a mathematical theory implied statements about history or about how many biological species there are, we would look on such mathematics with extreme suspicion, even if the statements about history or biology were correct. It is also worth noting the relationship between conservativeness and some of its neighbours. Necessary truth implies truth, but it also implies conservativeness. And both the latter imply consistency. Conservativeness is like necessary truth, without the truth. And according to Field, good mathematics need not be true, but it does need to be conservative.

Various objections are levelled at Field's programme, from claims about the implausibility of extending it to curved space-times (Urquhart 1990) and non-space-time theories (e.g., quantum mechanics) (Malament 1982),[4] through concerns about whether the nominalised science Field constructs has the theoretical virtues of its mathematical counterparts (Burgess 1983; Colyvan 2001) and whether a nominalist is entitled to be a substantivalist about space-time (Resnik 1985), to technical concerns over the logic Field relies on (first-order versus second-order logic and the account of modality) (Burgess and Rosen 1997). Despite what might seem like an overwhelming weight of criticism, it is important to recognise what Field's programme

---

[4] Malament (1982) argues that quantum mechanics, for one, is likely to resist nominalisation because of the central role infinite-dimensional Hilbert spaces play in the theory. Balaguer (1996; 1998), however, suggests a way of nominalising the Hilbert spaces in question.

achieves. It outlines a very attractive and uncompromising fictional account of mathematics, and one that does not shirk any of the major issues. There may be problems and it may be incomplete as it stands, but Field's philosophy of mathematics is not alone in either respect.

4.2 An easier route to nominalism?

While many philosophers are attracted to nominalism, the difficulties facing Field's approach lead them to explore other strategies. Another approach involves rejecting the claim that we need to take all the commitments of our best scientific theories seriously. In particular, this approach denies that mathematical entities are among the entities we need to be ontologically committed to – despite their indispensability to our best scientific theories. Several contemporary philosophers have been exploring such accounts, although some are not really nominalists (e.g., Balaguer 1998; Maddy 1997; Yablo 2005) in that they deny that there is a fact of the matter about whether mathematical entities exist. Still, there is a nominalist strategy in the vicinity of each of the positions in question, so I will treat them as nominalist strategies despite the defenders of some of these views declining to sign up to nominalism.

This kind of 'easy road' nominalism accepts the indispensability of mathematics to science, but denies that this gives us any reason to accept the existence of mathematical entities. The reasons for this denial vary, but a common suggestion is that, in some sense, scientific theories are about physical aspects of reality, and the positing of mathematical entities is merely a tool for expressing what is required. Consider a mixed scientific statement which invokes both mathematical entities and physical entities: 'There is a continuous function that maps from the space-time manifold to the real numbers such that certain conditions are satisfied.'

The nominalist takes this statement to be false (because there are no functions) but accepts that what is said about the world is true, namely, that the space-time manifold is as described. Note that the nominalist in question does not try to provide a mathematics-free translation of the mixed statement; that would involve a commitment to something like Field's programme. There are various ways to try to motivate the non-commitment to mathematical entities. One might argue, on independent grounds, that only causally active or spatiotemporally active entities exist or, more plausibly, one might try to argue that scientific practice itself does not commit one to the existence of mathematical entities (e.g., Maddy 1997). In the next sections we'll consider two such nominalist accounts: one due to Jody Azzouni and one due to Stephen Yablo.

4.2.1 Nominalism through thick and thin

Jody Azzouni has a very interesting and detailed nominalist strategy for the philosophy of mathematics. Azzouni's central idea is to distinguish between those scientific posits we ought to take to be real and those to be treated instrumentally. Azzouni is a realist about unobservable entities so he does not take the observable–unobservable distinction to mark the relevant cleavage here. He does admit, however, that there is something special about direct observation. With direct observation as the example of epistemic access par excellence, Azzouni then considers the important features of this kind of access.

He isolates four conditions direct observation satisfies (Azzouni 2004, pp. 129–36): *robustness*, *refinement*, *monitoring*, and *grounding*. Epistemic access satisfies *robustness* when the access does not depend on the expectations of the epistemic agent; for example, our theory about genetics might prove to be incorrect or might otherwise surprise us by outstripping our expectations. Epistemic access satisfies *refinement* when there are ways of adjusting and refining the epistemic access we have to the posit in question; for example, we can use more powerful microscopes to get a better look at micro-organisms. Epistemic access satisfies *monitoring* when we can track the posits in question by either detecting their behaviour through time or by exploring different aspects of the posits in question; for example, we can follow a particle via its track in a cloud chamber or we can walk around a mountain to view it from different aspects. Epistemic access satisfies *grounding* when particular properties of the entity in question can be invoked in order to explain how the epistemic access we have enables the discovery of those and other properties of the object; for example, we can identify the heart in a chest X-ray because its relative density means that it appears as a region of greater X-ray absorption and this, in turn, enables us to determine other properties of the heart, such as its size. As should be clear from some of these examples, direct observation is not the only kind of epistemic access to satisfy these conditions. These four conditions can be thought of as *generalisations* of features of typical direct observation. When we have access to unobservable particles such as alpha particles via a cloud chamber, we find that such access also satisfies these four conditions. Azzouni calls such access *thick epistemic access*. And as a generalisation of direct observation, it enjoys the privileged epistemic status of the latter.

Now contrast thick epistemic access with the kind of access we have via the role an entity plays in a scientific theory that enjoys the usual aesthetic virtues of simplicity, familiarity, and so on. Let us call such theoretically motivated access *thin epistemic access*, and contributing to the theoretical virtues of the theory in question. (Azzouni suggests that the latter amounts to an entity 'paying its Quinean rent'.) Entities accessed thinly may play indispensable roles in our best scientific theories, but intuitively they do not have the same kind of privileged status as entities accessed thickly. In addition to paying their Quinean rent, entities accessed thinly must also have a story in place explaining why they are not accessed thickly. For example, we might not be able to have thick access to the black hole at the centre of the Milky Way, but our very understanding of what a black hole is delivers a story of why we fail to have thick access (e.g., because black holes do not reflect or emit light). This 'excuse clause' turns out to do a lot of work for Azzouni.

The third kind of epistemic access Azzouni considers is *ultra-thin* access.[5] These we can think of as *mere posits*; they can be posited by anyone, at anytime, without any regard for reality. The posits of fiction are paradigmatic examples here. They need not play indispensable roles in our best scientific theories and they do not have excuse clauses for why they are not accessed thickly. Now we are in a position to draw the line between what's real and what's not. According to Azzouni, the thin–ultra-thin distinction is the crucial one: posits accessed either thickly or thinly are to be thought of as real. The ultra-thin, unsurprisingly, are not taken to be real, since they do not earn their keep. Azzouni explains how a thin posit can be demoted to ultra-thin, and the difference in attitude towards the two.

> The difference between thin posits and ultra-thin posits (which live free of charge) is striking. Should one of the former fail to pay its Quinean rent when due, should an alternative theory with different posits do better at

---

[5] It is the epistemic access to posits that is thick, thin, or ultra thin, although often it is convenient to speak of the posits themselves as thick, thin, or ultra-thin.

simplicity, familiarity, fecundity, and success under testing, then we have
a reason to deny that the thin posits, which are wedded to the earlier
theory, exist – thus, the eviction of centaurs, caloric fluid, ether, and their
ilk from the universe. (Azzouni, 2004, p. 129)

It is important to note that if a thin posit fails to deliver its excuse for why it
is not thick – even if its Quinean rent is paid – it will also find itself classified
as ultra-thin and thus evicted from Azzouni's ontology.

What we end up with is a way of distinguishing those portions of our
scientific theories that are taken to be real from those that are to be treated
instrumentally. Indeed, the cleavage produced is very similar to the causal–
acausal distinction. The thick posits are typically entities with which we
have causal contact, the thin are typically causal entities required by our
best scientific theories but with an excuse as to why we fail to have thick
(causal) access to them, and the ultra-thin are typically acausal entities. This
rough aligning of the causal–acausal cleavage and the real–instrumental
cleavage, presumably, is no accident. Earlier Azzouni (1997) toyed with
the idea of using the former as the means of distinguishing the real from
the instrumental. The problems associated with using a causal criterion,
however, are serious. Indeed, without independent motivation, such an
approach is simply question-begging. The beauty of Azzouni's thick and
thin epistemic access approach is that it does not seem to beg the question
against Platonism, and yet, according to Azzouni, it does rule against ontological commitment to abstract entities such as numbers. If all this were to
work, we would have a plausible easy road to nominalism.

4.2.2 Trouble in the tar pits

There are, however, some problems with this approach. First, note that the
thick, thin, and ultra-thin distinction is not sharp and yet it needs to be
in order to do the work required of it. The point is that epistemic access
can have the four crucial features – robustness, refinement, monitoring,
and grounding – in degrees. Take refinement, for example. Using a more
powerful optical telescope makes a big difference when looking at Saturn,
it makes less difference when looking at Alpha Centauri, and even the most
powerful optical telescopes make no difference at all when looking at a very
distant star whose presence is theoretically established (because, say, it is
having an influence on the motion of an observable binary partner). So do

we say that epistemic access to Saturn satisfies the refinement condition, epistemic access to Alpha Centauri *partially* satisfies it, and epistemic access to the distant star does not satisfy refinement at all? That seems reasonable enough. But now notice that although Azzouni will presumably accept all three posits as real, he will do so for three quite different reasons. He will accept Saturn as real because it is a thick posit. He will accept the star in the distant galaxy as real because it is thin – the defeasibility condition kicks in to explain why the access is not thick: the star in question is too far away. But what of Alpha Centauri? According to Azzouni, it will be a thick posit. But it won't be as thick as Saturn (or so we are supposing for the purpose of the example). Being less thick, it might plausibly require a partial excuse for not being as thick as we'd like. The excuse, of course, is that Alpha Centauri is a fair distance away (but not too far away for refinement to be impossible). But now this raises a serious question about the strength and nature of the defeasibility condition. It seems that some posits can be borderline thin. And given the above suggestion that the excuse clauses might come in varying strengths – some excuses are better than others – it may well be that the crucial thin–ultra-thin border is not sharp either. But perhaps that's OK. The lack of sharpness here does not need to correspond with a lack of sharpness in reality. It's not as though, according to Azzouni, there are partially existing objects. Rather, the vagueness in question corresponds to degrees of justification, and just means that there will be some objects which we should be agnostic about – the case for their existence is not compelling but neither is the case against their existence.

But worse still, there would seem to be clear cases of entities that do not fit Azzouni's tripartite classification. Those I have in mind enjoy the Quinean virtues but do not come equipped with an excuse for their lack of thick epistemic access. Let's call the access to such entities *very thin*. It is worth drawing attention to the importance of Azzouni's excuse clause concerning the lack of thick access. Recall that an entity is thin if we do not have thick access to it, but the entity in question pays its Quinean rent and has an excuse for its failure to support thick access. This excuse clause is important in order to avoid obvious counterexamples such as stars and planets outside our backward light cone. The latter are uncontroversially real but we cannot have thick access to them. They pay their Quinean rent and there is also a well-accepted story as to why we don't have thick access to such stars and planets: basically, they are too far away. But what if the excuse were not forthcoming? What should we say about very thin posits? Azzouni does not think that there are any, so does not tell us whether to count such posits as real or not. Such posits are thus confined to a kind of ontological purgatory: neither real nor unreal.

Some examples of such posits may help. Consider a 'gap' in the fossil record. This is a creature posited in order to make sense of the standard evolutionary story but with which we have no contact, let alone thick epistemic access. The crucial question is whether there is a story in place as to why we don't have thick access with such creatures. This is crucial because *with* a story, Azzouni is able to deliver the intuitively correct result that we are justified in taking such creatures to be real; *without* a story, such creatures turn out to be very thin and thereby sentenced to ontological purgatory. One excuse might simply be that such creatures are now extinct and so cannot be tracked. The fact that they existed in the past but are now extinct might be all that's required. But this seems too cheap. Surely we want a more substantial story about why such creatures never fell into tar pits or the like. But I take it (or at least we can suppose for the purpose of this example) that we don't have such a story. We are thus faced with two possibilities: (i) either these gaps in the fossil record are accessed very thinly and Azzouni gives us no advice about their ontological status, or (ii) they are accessed thinly because they come equipped with a fairly trivial and obvious story about why they are not accessed thickly.

Now to return to the case of interest: mathematical entities. These are not accessed thickly, on that (almost) everyone agrees. The question is whether they are accessed thinly, very thinly or ultra-thinly. Mathematical objects (at least prima facie) enjoy the Quinean virtues, so they are (at least prima facie) not ultra-thin. Whether they are thin or very thin depends on what can count as an excuse for not being accessed thickly. I suggest that mathematical objects, being acausal, have such an excuse. But is this excuse acceptable? Azzouni doesn't give us any guidance; he offers no systematic story about acceptable excuse clauses. Moreover, since the excuse clauses play a central role in Azzouni's account, independently of concerns about mathematical entities a well-motivated and detailed account of what passes for an excuse is required.

It is an interesting feature of Azzouni's account that many of our scientific (and historical) posits do not enjoy thick epistemic access: dinosaurs, Gondwanaland, the inflationary phases of the Big Bang, and Plato, to name a few. (Posits from past times are not able to be tracked and so cannot be accessed thickly.) It is, thus, clear that the issue of what counts as a permissible excuse for lack of thick access is crucial. If Azzouni is fairly liberal about such stories, then the excuse that mathematical entities are abstract may be acceptable. If he is too restrictive, Azzouni risks sliding into some form of scientific anti-realism – a kind of presentism, where only present objects can be thought to be real. In any case, whether the account of permissible excuses is liberal or restrictive, it needs to be independently motivated. After all, the ontological status of a large number of our theoretical posits will depend on the excuse clause. And without an independently motivated account of what excuses are admissible, we have no reason to take mathematical entities as unreal.[6]

4.3 Mathematics as metaphor

There's another position that's quite distinct from Platonism and nominalism – a kind of metaphysical nihilism whereby there is no fact of the matter about the existence of mathematical entities. In a sense, metaphysical nihilists are the common enemy of Platonists and nominalists alike. But they have one thing in common with nominalism: they're not Platonists. In any case, I include this position here in the chapter on nominalism because some of the nihilist strategies against mathematical realism can be (and have been) adapted by nominalists.[7] Several philosophers have advanced positions along these lines (Balaguer 1998; Maddy 1997; and Yablo 1998; 2005; 2009). Here I'll just very briefly sketch one such proposal due to Stephen Yablo.

Yablo likens mathematics in science to metaphor. He begins by noting that it would be a mistake to take metaphorical statements (and figurative language generally) to commit us to the objects apparently quantified

---

[6] In case it is not obvious, here is a place where I am unashamedly injecting my own views. The debate over whether Azzouni's nominalist account of mathematics succeeds is very much alive. The interested student will find a number of articles in the current literature both defending and criticising Azzouni's views – his position is attracting a great deal of attention. Moreover, this attention is well deserved. Azzouni's position is a major new development in the philosophy of mathematics.

[7] For example, Leng (2010) uses the nihilist criticisms of Quinean realism discussed in this section to mount a case for nominalism.

over in such language. Take, for example, the title of R. Dean Taylor's 1967 Motown single 'There's a Ghost in My House'.[8] Clearly we should not take the title's existential claim seriously; there is no reason to entertain the existence of ghosts, even if the title truly describes how it feels to live in a house after your partner has left. The metaphor in question is supposed to conjure up a suite of images. And like all interesting metaphors, the possible interpretations are never exhausted, and they require some interpretative work on the part of the reader. As Davidson (1978, p. 29) once suggested '[m]etaphor is the dreamwork of language, and like all dreamwork, its interpretation reflects as much on the interpreter as on the originator'.

Next, we note that we can invoke metaphors (and other forms of non-literal language) to truly describe actual situations. To use one of Ken Walton's examples, we can describe the Italian town of Crotone as being located on the arch of the Italian boot. Here the metaphor draws our attention to the similarity between the shape of Italy and a boot. We then engage in the pretence that Italy is a boot, and this pretence allows us to give (more or less) accurate information about the location of Crotone. Moreover, such figurative language is present in our scientific discourse – average stars and so on – and such uses are arguably ineliminable. Yablo then argues that there is no clear boundary between the portions of scientific discourse intended literally and those which are merely metaphorical. This leads to a serious problem for the Quinean. Clearly we should only read off our ontological commitments from literal parts of our scientific theories, but if these theories are shot through with figurative language, we need to be able to separate the literal from the figurative before we can begin ontology. But here's the kicker: according to Yablo, there is no way of separating the literal from the figurative.

> To determine our commitments, we need to be able to ferret out all
> traces of non-literality in our assertions. If there is no feasible
> project of doing that, then there is no feasible project of Quinean
> ontology. (Yablo 1998, p. 233)

Yablo considers only descriptive uses of language in science – language intended to *describe* the state of some system. He does not consider uses of scientific language intended to *explain* why some system is in a particular

---

[8] Written by Brian Holland, Lamont Dozier, Edward Holland, and R. Dean Taylor.

state. Does this make a difference? I think it does. It may well be right that metaphorical language intended only to describe need not carry onto-logical commitment, or at least, it need not carry the obvious, literal ontological commitments ('ghosts' and the like), but it's not clear that lan-guage intended to deliver explanations can be thought to be free of such commitments.

So let's grant that metaphorical language (and figurative language gener-ally) can be used for purposes of true description, as Walton and Yablo rather convincingly argue. The important question for our purposes is whether figurative language can be explanatory. Take, for example, the sentence 'The owner of the motel is unhinged.' This sentence invokes a metaphor to describes certain psychological features of the motel owner, but it might also be thought to *explain* why you might be reluctant to stay at the motel in question. And as we've already seen, there is no need to take the ontological commitments of the metaphorical language seriously – no need to expect the motel owner to consist, in part, of broken hinges or to make noises like a broken shutter. But wait! How can a metaphor, invoking non-existent entities, explain? Answer: the explanation of the metaphor stands proxy for some further *real* explanation. The real explanation is that we are reluctant to stay at the motel in question because we fear that the motel owner can-not be relied upon to act appropriately (where this, in turn, stands proxy for a more complicated story about the motel owner's cognitive states and capacities). The important point to note here is that, to the extent that the metaphor is explanatory, any explanation delivered by the metaphor is really just standing proxy for another, more complicated explanation. In any case, the ontological baggage of the metaphorical explanation, the hinges, for example, *do not play any essential role in the explanation*. That's the thought, at least.

This raises the question of whether there are cases where fictional enti-ties invoked by a metaphor carry some of the explanatory load. Yablo argues for a number of different ways in which metaphors are essential, but one he doesn't consider is *metaphors essential for explanation*. Suppose you want to explain why someone is feeling anxious and depressed after their partner has left, and you say that there's a ghost in their house. Clearly this won't do. For the metaphor to function as an explanation, either there must lit-erally be a ghost in their house, in which case it's not a metaphor at all, or it's not an explanation, because it relies on non-existent entities. It seems that metaphors can carry explanations only when the metaphor in question stands proxy for some non-metaphorical explanation. It is hard to see how there could be metaphors essential to explanation. At least Yablo has not established that there are any such cases.

So my suggestion here is that when some piece of language is delivering an explanation, either that piece of language must be interpreted literally or the non-literal reading of the language in question stands proxy for the real explanation. Moreover, in the latter case, this real explanation must be understood to be conveyed by the metaphor in question.[9] If this is right, we have the makings of at least a partial response to Yablo's challenge to mark the boundary between the literally true parts of our theory and the figurative: whenever we have an explanation that is not simply a metaphor standing proxy for some other real explanation, we ought to treat the language in question as literal and thus as being ontologically committing. It remains to show that there are cases in scientific discourse where mathematics features in explanations. If we can show this, then anyone tempted by Yablo's metaphysical nihilism will have one of two options: (i) provide suitable and well-understood translations of the mathematical explanations in question; or (ii) show why the alleged explanations in question are not really explanations at all. If neither of these is possible, we have good reason to accept the explanations at face value and to take their ontological commitments seriously.[10]

The issue of mathematical explanations has arisen several times in the last couple of chapters. In the next chapter we turn to a more thorough treatment of this topic.

Discussion questions

1. In line with our (rough) definition of *conservativeness* given in this and the last chapter, the conservativeness of mathematics can be defined more

---

[9] So rich metaphors, where the possible interpretations (and even *intended* interpretations) are not exhausted by any convenient translation, do not seem capable of delivering genuine explanation.

[10] Again I'm taking a stand on substantial issues here: issues that – at the time of writing – are considered to be wide open. I encourage you to think critically about both my and Yablo's position here. Whether you agree with Yablo or not, his position is interesting, novel, and an important development in the debate over mathematical realism and nominalism.

formally as: a mathematical theory M is said to be conservative if, for any body of nominalistic assertions S and any particular nominalistic assertion C, then C is not a consequence of $M + S$ unless it is a consequence of S. But even this is not quite right. Consider a case where S contains the sentence 'there are no mathematical entities'. This would render $M + S$ inconsistent. How would this trivialise the definition? How would you refine the definitions to avoid such difficulties?

2. One of Field's formal proofs of the conservativeness of mathematics uses set theory. Is it legitimate for him to use mathematics in the process of showing that mathematics is dispensable?

3. As mentioned, one of Field's motivations is to exclude extrinsic entities and seek intrinsic explanations. Field says, 'one wants to be able to explain the behaviour of the physical system *in terms of the intrinsic features of that system*, without invoking extrinsic entities (whether non-mathematical or mathematical) whose properties are irrelevant to the behaviour of the system being explained' (Field 1989, p. 193). But Field also expresses sympathy towards a unification account of explanation which involves reducing the number of independent phenomena, often by making some surprising connections (Field 1993, p. 295). (See section 5.1 below for more on this account of explanation.) Is there a tension here? Can explanation both unify and be intrinsic?

4. Let's suppose that fictionalists are right and the so-called truths of mathematics are only true according to the story of mathematics. So that '7 is prime' is false, but it is true in the story of mathematics. It is tempting to introduce an 'in the story of mathematics' operator into our language so that we can recover truth simpliciter. 'In the story of mathematics 7 is prime' is thus simply true. Now consider the sentence 'Erdös loved numbers.' Try to insert the 'in the story of mathematics' operator in such a way that this sentence comes out true simpliciter. Or harder still: 'Erdös loved numbers more than he loved anything else – real or fictional.'

5. How might Azzouni respond to the challenge of spelling out what counts as a legitimate excuse for failing to provide thick access? Do you think 'being without causal powers' should count as a legitimate excuse?

6. Does science really use figurative language in fully-developed theories? Is the task of disentangling the literal from the figurative elements of scientific discourse as hopeless as Yablo suggests?

7. Can there be purely metaphorical explanations? That is, can there be explanations that employ metaphorical language and such explanations are not just standing proxy for the real explanation?

Recommended further reading

Field (1980) provides a fairly technical account of the most influential fictionalist programme in the philosophy of mathematics. Balaguer (2009) presents an alternative fictionalist strategy. A good overview of fictionalism in metaphysics can be found in Eklund (2007). Burgess and Rosen (1997) provide a very good and critical overview of nominalist strategies, generally, and Burgess (1983) offers some criticisms of nominalism. Azzouni's 'easy road' strategy can be found in Azzouni (2004) and also in a number of papers such as Azzouni (1997), and critically discussed in Colyvan (2005). For Yablo's criticisms of the Quinean approach to metaphysics see his (1998; 2005). Leng (2010) puts this to work in developing a nominalist philosophy of mathematics. Another major 'easy road' strategy can be found in Melia (2000).

Azzouni, J. 1997. 'Applied Mathematics, Existential Commitment and the Quine–Putnam Indispensability Thesis', *Philosophia Mathematica*, 5(2): 193–227.

  2004. *Deflating Existential Consequence: A Case for Nominalism*, Oxford University Press.

Balaguer, M. 2009. 'Fictionalism, Theft, and the Story of Mathematics', *Philosophia Mathematica*, 17: 131–62.

Burgess, J. 1983. 'Why I Am Not a Nominalist', *Notre Dame Journal of Formal Logic*, 24(1): 93–105.

Burgess, J. P. and Rosen, G. A. 1997. *A Subject with No Object*, Oxford: Clarendon Press.

Colyvan, M. 2005. 'Ontological Independence as the Mark of the Real', *Philosophia Mathematica*, 13(2): 216–25.

Eklund, M. 2007. 'Fictionalism', in E. N. Zalta (ed.), *The Stanford Encyclopedia of Philosophy* (Summer 2007 edn), http://plato.stanford.edu/archives/sum2007/entries/fictionalism/.

Field, H. 1980. *Science without Numbers*, Oxford: Blackwell.

Leng, M. 2010. *Mathematics and Reality*, Oxford University Press.

Melia, J. 2000. 'Weaseling Away the Indispensability Argument', *Mind*, 109: 453–79.

Yablo, S. 1998. 'Does Ontology Rest on a Mistake?', *Proceedings of the Aristotelian Society, Supplementary Volumes*, 72: 229–61.

  2005. 'The Myth of the Seven', in M. Kalderon (ed.), *Fictionalism in Metaphysics*, Oxford: Clarendon Press, pp. 88–115.

5 Mathematical explanation

In mathematics you don't understand things. You just get used to them.

John von Neumann (1903–57)[1]

Mathematical explanation is a hot topic in current work in the philosophy of mathematics. We have already seen one reason for this: the close connection between the indispensability argument for mathematical realism and the scientific realist's reliance on inference to the best explanation. This connection is even tighter if it can be established that there are mathematical explanations of empirical phenomena. As a result, a great deal of recent work on realism–anti-realism issues in mathematics has focused on mathematical explanations in science. Irrespective of such issues, the question of mathematical explanation is important in its own right and deserves closer attention.

We start by making a distinction between two different senses of mathematical explanation. The first we call *intra-mathematical* explanations. These are mathematical explanations of mathematical facts. Such explanations can take the form of an explanatory proof – a proof that tells us why the theorem in question is true – or perhaps a recasting of the mathematical fact in question in terms of another area of mathematics. There is also the issue of whether mathematics can explain empirical facts. Call this *extra-mathematical explanation*. A full account of mathematical explanation will provide both a philosophically satisfying account of intra-mathematical explanation and an account that coheres with our account of explanation elsewhere in science.

---

[1] Attributed to von Neumann, as quoted in G. Zukav, *The Dancing Wu Li Masters: An Overview of the New Physics*, New York: William Morrow, 1979, p. 208.

5.1 Theories of explanation

At the most basic level, an explanation is a story that makes something that is initially puzzling less puzzling; an explanation reduces mystery. Often explanations are offered as replies to 'why' questions: 'Why did the species become extinct?'; 'A very efficient predator was introduced.' 'Why do pulsars spin so fast?'; 'They collapse under gravitational forces and as a result of the conservation of angular momentum, there is an increase in angular velocity.'

There are a number of different philosophical theories of scientific explanation. Most, however, are non-starters as accounts of mathematical explanation. For example, according to one popular line, an explanation involves identifying relevant causes of the explanandum (that which is to be explained). So the explanation of the recent eruptions of the Icelandic volcano Eyjafjallajökull would involve an account of the build-up of magma and gases and their eventual release, after sufficient pressure was reached. It is clear that whatever virtues such causal/mechanical accounts of explanation enjoy, these accounts are of no use for mathematical explanations. One can hardly appeal to *causes* to explain why the Fundamental Theorem of algebra holds or why Pythagoras' Theorem holds.

Other accounts of explanation involve probability theory and statistical relevance. For example, according to one account, an explanation of some event E is given in terms of another event C such that $P(E|C) > P(E)$. The intuitive idea here is that the event C raises the probability of E, and as such, C may serve to (at least in part) explain E. Think of a case where E is initially very unlikely, but learning of some other event makes E rather likely. Again, such statistical accounts of explanation clearly have something going for them, but they are no use for mathematical explanation. The usual understanding of mathematics is that it is a body of necessary truths. That is, a mathematical statement is either necessarily true or necessarily false. This, in turn, means that all the probabilities in question are either 0 or 1. So for the probabilistic account of explanation just sketched to work here, mathematical explanations would need to raise the probability of some mathematical explanandum from 0 to 1. But this is not possible, at least it's not possible on standard accounts of probability theory where if $P(E) = 0$, there is no C such that $P(E|C) > 0$, let alone $P(E|C) = 1$.[2] In

---

[2] Oddly, if $P(E) = 0$, even $P(E|E) \neq 1$. $P(E|E)$ is not defined.

any case, according to a common-sense view of mathematics, mathematical statements always have probability 1 or probability 0; the whole notion of probability raising seems to be on the wrong track.

Finally, an account of explanation that does seem to have some legs when it comes to mathematics is the unification account.[3] According to this account, explanation is the business of unifying a collection of perhaps disparate facts under a single overarching theory. The classic examples of unification in science include Newton's theory of gravitation, which unifies such phenomena as the tides and celestial mechanics. It also seems plausible that Newton's theory – via the unification in question – also explains the tides and various details of celestial mechanics. One point worth noting about this account of explanation is that it does not require that the overarching theory is epistemically more secure or more fundamental. Indeed, in the case of Newtonian gravitational theory, the overarching theory was rather contentious at the time, in that it invoked a notion of gravitation acting at a distance. But still gravitation could be invoked to explain the tides, for example, because of the unification – not because gravitation is better understood than the tides. At least, that's how explanation is supposed to work on the unification account.

The unification account of explanation faces a number of challenges. These included some alleged counterexamples and the problem of spelling out crucial notions, such as 'unification', in anything other than metaphorical terms. Indeed, it's fair to say that the problems facing the unification account are considered serious enough to make it something of an outsider in current discussions of scientific explanation. Be that as it may, this account strikes me as very well placed to make sense of mathematical explanation. I'll have more to say about this in the next section.

5.2 Intra-mathematical explanation

We need to take seriously mathematicians' claims that there is explanation in mathematics. For instance, Fields medalist[4] Timothy Gowers

---

[3] An older, now largely abandoned account of explanation known as the deductive-nomological (or D-N) account also has some credibility as an account of mathematical explanation.

[4] The Fields medal is perhaps the highest honour in mathematics. It is awarded every four years to a mathematician of 40 or younger 'to recognize outstanding mathematical

and Michael Neilson (2009, p. 879) point out that:

> [F]or mathematicians, proofs are more than guarantees of truth: they are valued for their explanatory power, and a new proof of a theorem can provide crucial insights.

Not all philosophers take such claims seriously. But those who would dismiss mathematical explanation are on very shaky ground. Some, for example, reject mathematical explanation because it is not accommodated by their preferred philosophical account of explantation. To my way of thinking, that's getting the order of things the wrong way around. The job of philosophers of science and mathematics is to help make sense of, and contribute to, science and mathematics *as practised*. The role of philosophers is not to overrule the pronouncements of mathematics and science on philosophical grounds – at least not pronouncements on matters of mathematics and science. In any case, there does not seem to be a completely satisfactory philosophical theory of scientific explanation, so to rule out mathematical explanation on the grounds of an unsatisfactory philosophical theory is simply poor methodology.

Instead, I propose we take mathematical explanation seriously, since it is taken seriously by mathematicians. If mathematical explanation does not sit well with some philosophical theory of explanation, so much the worse for the philosophical theory of explanation. It may be that mathematicians are mistaken when they suggest that there is explanation in mathematics, but that needs to be argued for on independent grounds. It cannot be taken as the starting point for discussion of explanation. And if there is no mathematical explanation, that cannot be taken as having been established by a failure on the part of philosophy to deliver an account of explanation that accommodates mathematical explanation. OK, that's the end of the sermon on methodology!

In order to make progress on the topic of mathematical explanation, we first need to get a feel for what mathematicians have in mind when they talk of explanations in mathematics. So let's consider a couple of ways explanations seem to arise in mathematics and some examples of each.

achievement for existing work and for the promise of future achievement' (taken from the Fields medal website: www.mathunion.org/general/prizes/fields/details/, accessed 26 December 2010).

5.2.1 Explanatory and non-explanatory proofs

Mathematicians value explanatory proofs and contrast these with unexplanatory proofs. But it is difficult to find a great deal of agreement in the mathematical literature on which proofs are explanatory and which are not. This is not because such issues are irrelevant to mathematical practice, it's just that some things don't find their way into the final published articles. After all, to claim that one proof of a particular theorem is more explanatory than some other proof of the same theorem invites the question: on what basis is this claim being made? Without a theory of mathematical explanation, it is very difficult to answer such questions. It's better to leave explanatory comparisons out of published work and let others decide for themselves. But, still, one encounters talk of explanatory virtues of theorems in the maths classroom and in other more informal settings.

One kind of proof that is generally thought to be unexplanatory is the *reductio* proof. Recall that these are proofs that proceed by assuming the negation of the proposition to be proved and deriving a contradiction from it. So, for example, consider Euclid's proof (*c.* 300 BCE) that there are infinitely many prime numbers.[5]

Theorem 2 (Euclid). *There are infinitely many prime numbers.*

Proof. Assume that there is a largest prime. Call it p_ℓ. Now consider the number one greater than the product of all the primes: $n = 2 \times 3 \times 5 \times ... \times p_\ell + 1$. Either n is a product of primes or it is a prime larger than p_ℓ. If the latter, we have a contradiction, since by assumption p_ℓ is the largest prime. So n must be a product of primes. But now we show that at least one of the primes must be greater than p_ℓ. If n is a product of primes and has no prime factors greater than p_ℓ, then one of its factors, q, must be in the sequence $2, 3, 5, ..., p_\ell$. It therefore divides the product $2 \times 5 \times ... \times p_\ell$. However, since it is a factor of n, it also divides n. But a number which divides two numbers also divides their difference, so q must also divide $n - (2 \times 3 \times 5 \times ... \times p_\ell) = (2 \times 3 \times 5 \times ... \times p_\ell + 1) - (2 \times 3 \times 5 \times ... \times p_\ell) = 1$.

---

[5] Recall that a natural number is *prime* if it is greater than 1 with no divisors apart from itself and 1. (A natural number greater than 1 is *composite*, otherwise 1 is neither prime nor composite.)

However, no prime divides 1 so q is not a prime in the sequence $2, 3, 5, ...p_\ell$. It follows that if n is composite, it has at least one factor greater than p_ℓ. This is a contradiction. There is thus no largest prime; there are infinitely many primes.[6]   □

What this proof establishes is that there must be infinitely many prime numbers because otherwise we'd get a contradiction. Arguably, this does not help us see *why* there are infinitely many prime numbers. However, I resist suggesting that all *reductio* proofs are unexplanatory. Some *reductio* proofs are very closely related to other forms of proof and in such cases either both are explanatory or neither is. Indeed, the proof above is easily converted into a direct proof. Further, it seems to me that *reductio* proofs can be explanatory. Consider a *reductio* proof of the fact that 2 is the only even prime. We assume, for the purposes of the *reductio*, that there is another even prime, $p > 2$. Since p is even, it can be divided by 2 so it can be written as $p = 2q$ for some integer $q > 1$. But this means that p is composite, contradicting the original assumption that p is prime. But this proof, I take it, *is* explanatory. It tells us why 2 is the only even prime: every other even number is divisible by 2 and hence cannot be prime. The explanation here is trivial, but still it's an explanation. So it seems that there is nothing about *reductio* proofs in general that makes them unexplanatory. Rather, it depends on the details of the proof in the particular case. The proof may proceed via the explanatorily relevant concepts, in which case it will be explanatory. But in many cases, *reductio* proofs do not do this – they just crank out a contradiction by any means available, and in such cases they are not explanatory.

Another style of proof often cited as unexplanatory are brute-force methods such as proof by cases (or proof by exhaustion). Here the proof proceeds by recognising that there are a number of different cases which exhaust the possibilities. We then prove that the result in question holds for each of the cases. An example will help. One proof of Rolle's Theorem in calculus proceeds by cases.

---

6 The famous English number theorist G. H. Hardy (1877–1947) said that this proof is 'as fresh and significant as when it was discovered' and that 'two thousand years have not written a wrinkle on [it]' (Hardy 1967, p. 92).

Theorem 3 (Rolle's). *Let f be a function which is continuous on the closed interval $[a, b]$ and is differentiable on the open interval (a, b). Suppose $f(a) = f(b) = 0$. Then there is a real number c such that $a < c < b$ and $f'(c) = 0$.*

Proof. We divide the situation into three cases, which are exhaustive: (i) f is never positive or negative on the interval (a, b); (ii) f is positive somewhere on the interval (a, b); or (iii) f is negative somewhere on the interval (a, b). Case (i): $f(x) = 0$, for all x, which means that $f'(x) = 0$ for all $x \in (a, b)$. Case (ii): a continuous function on a closed interval has a maximum in the interval in question. Since f is positive somewhere in (a, b), the maximum value of f must be positive. Since $f(a) = f(b) = 0$, f takes its maximum value at some point c in the open interval (a, b). Since $a < c < b$, f is differentiable at c and since c is a maximum of f, $f'(c) = 0$. Case (iii): a continuous function on a closed interval has a minimum in the interval in question. Since f is negative somewhere in (a, b), the minimum value of f must be negative. Since $f(a) = f(b) = 0$, f takes its minimum value at some point c in the open interval (a, b). Since $a < c < b$, f is differentiable at c and since c is a minimum of f, $f'(c) = 0$.   □

Proofs such as this lack unity. Different reasons are often offered in the different cases, and it looks as though the theorem itself holds merely by accident.[7] What we would like is a proof that offers the same reason in each case; that would provide an explanation of the theorem in question. But then such a proof would not need to be done by cases. This is at least part of the reason why mathematicians find proof by cases less satisfying than other styles of proof.

Finally, a more controversial example: proof by *mathematical induction*. Mathematical induction, although sharing some characteristics of scientific induction, should not be confused with it.[8] Unlike scientific induction, mathematical induction is a deductive inference. Mathematical induction

---

[7] See Baker (2009a) and Lange (2010) for more on mathematical accidents and coincidences.

[8] Scientific induction is a non-deductive inference from a finite (or in some cases infinite) number of observed instances that are thus and so, to the general claim that *all* cases are thus and so (or at least that some unobserved cases are thus and so). For example, the inference from the sun has always risen in the past to the conclusion that the sun will always rise in the future (or that it will at least rise tomorrow), is a scientific inductive inference.

proofs have the following form: (i) show that some statement holds for a base case (for a particular natural number); (ii) show that if the statement holds for some arbitrary number, then it holds for that number's successor; then (iii) it can be concluded that the statement holds for all natural numbers greater than or equal to the base case. As usual, an example will help. Here we show that the sum of natural numbers up to n is $n(n + 1)/2$.

Theorem 4 (Sum of Natural Numbers Formula). $\sum_{i=0}^{n} i = n(n + 1)/2$.

Proof. Note that $0 = 0 \cdot (0 + 1)/2$, so it holds for the base case of $n = 0$. Next we assume that (1) $\sum_{i=0}^{k} i = k(k + 1)/2$ and we need to show that (2) $\sum_{i=0}^{k+1} i = (k + 1)(k + 2)/2$. But $\sum_{i=0}^{k+1} i = \sum_{i=0}^{k} i + (k + 1) = k(k + 1)/2 + (k + 1) = k(k + 1)/2 + (2k + 2)/2 = (k^2 + 3k + 2)/2 = (k + 1)(k + 2)/2$ (as required).[9]   □

It is hard to make much progress on the issue of the explanatoriness of mathematical induction proofs, precisely because there is such disagreement about whether such proofs are explanatory. On the one hand, proofs such as the one just presented seem unexplanatory because the proof feels mechanical, and the only way the content of the theorem enters into our reckoning is via some rather trivial algebraic manipulations. Indeed, all mathematical induction proofs have the same structure (outlined above), and this structure does not depend on the content of the theorem. On the other hand, mathematical induction proofs such as this work because of the structure of the natural numbers: each number has a unique successor, and all but 0 have a unique predecessor.[10] Indeed, the fact that mathematical induction holds of the natural numbers is in some ways a characterising feature of them. So, it might be argued that any proof by induction is in fact revealing the explanation of the theorem in question, namely, that it holds by virtue of the structure of the natural numbers. This is not the most satisfying explanation, but perhaps that's all there is to it and this explanation is faithfully delivered by mathematical induction proofs.

But in line with what I was suggesting earlier, I'm reluctant to attribute explanatoriness or lack thereof to styles of proof. Indeed, it is important to

---

[9] Note the crucial use of the inductive assumption (1) and how an opening was made for its use by splitting $\sum_{i=0}^{k+1} i$ into $\sum_{i=0}^{k} i$ and $(k + 1)$.

[10] Of course, in the integers, 0 has -1 as its predecessor, but in the natural numbers 0 has no predecessor.

remember that not all mathematical induction proofs are concerned with simple summation facts about the natural numbers, as in the classic example above. In any case, many mathematical proofs proceed by conditional proof. These are proofs that take some assumptions, A_j, then from these show that some further statement C holds. The conclusion is that the conditional holds: if A_j then C. It would be very odd if simply being a conditional proof were enough to make the proof in question either explanatory or not. I suspect that the feel of unexplanatoriness about inductive proofs comes from considering only simple examples. Although examples such as the one I presented above are the best known, mathematical induction proofs are much richer. One possibility is that, as I suggested, there is not much by way of explanation to be had in these simple cases, but that's not the fault of the style of proof. Perhaps in more complex cases, there is more to do in establishing the inductive clause and that work might deliver explanations. Of course this is mere speculation, but before we make pronouncements either way about the structure of mathematical induction proofs, we should consider a wide variety of such proofs.[11]

More promising than making pronouncements about explanatoriness or lack thereof based on the structure of the proofs in question, is to take the details of the proof more seriously and look there for guidance. One way to do this is to look for relevance. Are the mathematical concepts invoked in the course of the proof relevant to the content of the theorem being proven? Now, 'relevance' sounds like a rather subjective piece of jargon that is no better understood than 'explanation', but this is not so. It turns out that a great deal of work has been done on logics that attempt to capture the notion of relevance – so-called *relevant logics*. Thanks to this work in logic, the notion of relevance can be made rigorous and spelled out in a systematic way.

To get a feel for the idea of relevance, consider the following argument, which is valid in classical logic: from the assumption that I'm alive, it follows that if I'm dead then I'm alive. Here we assume that the conditional in

[11] And these should include transfinite inductions. These are inductive proofs that proceed as above but include an extra step to show that (iia) if the statement holds for all cases less than some limit ordinal, the statement also holds for the limit ordinal. Such proofs may give insights into what's going on at limit ordinals (or so it seems to me, at least). See Baker (2010) and Lange (2009) for an interesting recent debate over the explanatoriness of proofs by mathematical induction.

question is the material conditional of classical logic. There are many ways to object to this argument. One is to protest that the material conditional of classical logic is a poor formal counterpart of the English 'if ... then ...' locution. Be that as it may, we can do more to diagnose the problem here. According to one line of thought the problem is that the material conditional does not respect relevance. Note that the assumption of the proof is not relevant to the conclusion. All that the conclusion requires is for the antecedent of the conditional to be false. We can just as easily prove that 'if $2 + 2 = 5$, then Fermat's Last Theorem', 'if Australia is uninhabited, then there are only finitely many prime numbers', and 'if Elvis Presley is still alive, Oswald killed John F. Kennedy'. In each case there is a disconnection between the antecedent of the conditional and the consequent. Compare these conditionals with ones such as 'if my pulse is strong and I have brain activity, then I'm alive' and 'if the Warren Report is correct, then Oswald killed John F. Kennedy'. In these latter conditionals the antecedent is relevant to the consequent. We need to be able to distinguish between such cases. Relevant logics are logics designed to recognise and respect such distinctions.

The central idea of relevant logic is to place a further condition on proofs: as well as being classically valid, the premises must also be relevant to the conclusion. (As I said earlier, the notion of relevance is technical and can be rigorously spelled out, although we needn't bother with the details here.) I am not suggesting that mathematics needs to abandon classical logic and replace it with relevant logic. Rather, I am suggesting that perhaps the distinction between an unexplanatory proof and an explanatory proof lies in the distinction between a merely classically valid proof and a relevantly valid proof (respectively).[12]

It is also worth noting that the unification account of explanation is well placed for making sense of explanatory and unexplanatory proofs in mathematics. Recall that the unification account of explanation sees explanation as occurring when disparate phenomena are unified under an overarching theory. For example, proof by cases typically lacks the unity required for explanation to be delivered. In other proofs it will depend on the content, not in terms of relevance, as suggested earlier, but in terms of illuminating

---

[12] See Mares (2009) for a very good introduction to relevant logics and Mares (2004) for more detail, philosophical motivation, and interpretation of these logics.

connections between different branches of mathematics that in some cases are yet to be unified. For example, a proof of a fact about the natural numbers that proceeds via the complex numbers might be seen to be forging a connection between the theory of the natural numbers and the theory of the complex numbers. A case can be made, at least, that on the unification account of explanation, such a proof would be explanatory. After all, a derivation of the behaviour of the ocean tides from the orbit of the moon in Newtonian mechanics is a canonical example of an good explanation on the unification account. In the tides case, we forge a link between two apparently disparate phenomena – ocean tides and celestial mechanics – and thereby shed light on both. Arguably, proofs in mathematics that invoke mathematical results from elsewhere are also shedding light on both areas and thus offering explanations.

Needless to say, the discussion here is rather speculative. Philosophers of mathematics are yet to provide a workable account of explanation for mathematical proofs, but both relevance and unification seem promising places to start. To make progress here we first need a clearer picture of which proofs mathematicians find explanatory and which they do not find explanatory.

5.2.2 Explanatory bridges

While most of the discussion of intra-mathematical explanation has focused on proofs, these are not the only loci of explanation. If the unification account of explanation is right, we should also expect to see explanation in reductions of one theory to another and in various generalisations of a theory. Of course, I am not suggesting that we should accept the unification account of explanation, but it does seem to imply that explanation will be found in places other than proofs. It is thus worth briefly looking at some examples to see if a case can be made for explanation existing in reductions and generalisations.

Let's start with a familiar example from set theory. Set theory has long been thought to be a foundation for the rest of mathematics, in the sense that it is the most abstract of the mathematical theories and that all other mathematical theories can be modelled in set theory.[13] For a start we've

---

[13] We set aside the even more general and abstract category theory.

learned that we can model the natural numbers with the von Neumann ordinals thus: $0 = \emptyset$; $1 = \{\emptyset\}$; $2 = \{\emptyset, \{\emptyset\}\}$; $3 = \{\emptyset, \{\emptyset\}, \{\emptyset, \{\emptyset\}\}\}$ ..., where the successor relation $S(x)$ is simply $x \cup \{x\}$. We then note that ordered pairs of natural numbers (a, b) can be modelled set-theoretically thus: $\{\{a\}, \{a, b\}\}$. Since rational numbers are just (equivalence classes of) ordered pairs of natural numbers, we can 'build' them out of the above construction for the natural numbers and our set-theoretic ordered-pair construction. Real numbers are technically a bit harder since they involve identifying each real number with a sequence of rational numbers, but this too can be modelled set-theoretically. From there we get complex numbers as ordered pairs of real numbers, real-valued functions as sets of ordered pairs of real numbers, and so on. The details are non-trivial, and I don't mean to suggest that any of this is easy or obvious. Nor am I suggesting that with these constructions in place, all we need is set theory so we can dispense with real analysis, complex analysis, and the like. As the first step should make clear, the notation for even small natural numbers such as 7 is unwieldy. And just imagine what a simple real-valued function such as a parabola looks like set-theoretically!

The point of such reductions is not to do away with everything except set theory. What is the point then? We might invoke Occam's razor at this juncture and argue that the reduction in question shows that all of mathematics is really just set theory. We might still permit the usual non-set-theoretic notation for the sake of convenience. On this view, we take the constructions in question to support a kind of ontological reduction of mathematics to set theory. Quine, for instance, held this view. But this is not forced upon us. Especially with our present interests in explanation to the fore, we might think that it is a very interesting fact about set theory that it can model much of the rest of mathematics in this way. We might take the construction in question to shed light on the relationship between set theory and the rest of mathematics in much the same way a computer model of a physical system is supposed to shed light on the target physical system (and sometimes also to shed light on the computer model itself). But it is not clear that such constructions do give us a better understanding of the natural numbers, for example. On the one hand, we understand pretty well what the number 3 is, right? In any case, seeing it as $\{\emptyset, \{\emptyset\}, \{\emptyset, \{\emptyset\}\}\}$ and the successor of $\{\emptyset, \{\emptyset\}\}$ doesn't give us any further insights. Or does it? Perhaps seeing the otherwise primitive successor function as just an application of set-theoretic union helps in the sense that we now have only one primitive instead of two. Seeing the complex numbers as ordered pairs of real numbers and thinking of them geometrically represented on the Argand plane, is a great help in coming to grips with the complex numbers.[14] Historically, such representations were important in demystifying the complex numbers. I'm inclined to think that there is explanation in such representations, but I admit that this is controversial.

Next consider generalisations in mathematics. There are at least two different kinds of generalisation. The first is extending a system to go beyond what it was originally set up for. The second involves abstracting away from some details in order to capture similarities between different systems. An example of extending a mathematical system is the move from the natural numbers, to the integers (introducing the negative numbers), then to the rationals, to the reals, and the complex numbers. It is important to note that each time such an extension is made, decisions need to be made about how to extend the familiar operations such as $+, -, \times$, and exponentiation so that they are extensionally equivalent on the old domain but give sensible answers on the new domain. The extensions must also give sensible answers on mixtures of the old and new domains. Even extending $+, -$, and $\times$ to the integers, rationals, and reals is non-trivial, but familiarity with the operations in the extended domains makes it seem more straightforward than it is. But consider the less intuitive exponentiation.

To start with, in the natural numbers it is just a bit of cute notation so that we can succinctly write $a \times a$ (b times) as a^b, where both a and b are natural numbers and $b > 1$. Let's consider $b = 1$ and $b = 0$. $a^1 = a$, for all a is an obvious way to deal with $b = 1$, but what about a^0? We define that to be 1, for all a. That's not obvious and is decided by how well that stipulation works in various mathematical applications. For example, for all $b > 2$ and $a \neq 0$, $a^{b-1} = a^b/a$. It seems reasonable to let this hold for $b = 2$ and $b = 1$ as well, which gives us $a^1 = a$ and $a^0 = 1$, respectively.

[14] The Argand plane (named after Jean-Robert Argand (1768–1822)) is like the Cartesian plane, but with the purely imaginary numbers ..., $-2i$, $-i$, 0, i, $2i$, $3i$, ... along the y-axis and the real numbers along the x-axis. Any complex number $x + yi$ can thus be represented as a unique point in the Argand plane.

Now we extend the domain to the integers. Integers raised to natural number powers can be defined in the obvious way (but they do need to be defined, for the original definition applied only to natural numbers). But to make sense of an integer raised to the power of a negative integer, we need only consider the above argument and let the equation $a^{b-1} = a^b/a$ hold for $b = 0$, which gives us $a^{-1} = 1/a$, and by extension $a^{-b} = 1/a^b$. As natural as these definitions may seem, choices had to be made and these choices were justified by the way the definitions in question cohere with various mathematical considerations.

Skip through another couple of non-trivial domain extensions with the rationals (making sense, for example, of $a^{1/2}$) and the reals (making sense of, for example, a^π) and now think about what it means to raise a complex number to the power of another complex number. This is where things get really interesting. It turns out that complex exponentiation needs to be spelled out in terms of (complex) trigonometric functions. I won't go into all the details of the general case here, but the case of raising the number e to a complex power will illustrate the idea: $e^{x+yi} = e^x(\cos y + i \sin y)$. This is Euler's formula and provides deep insights into complex analysis, trigonometry, and exponentiation. Complex exponentiation is not merely an arbitrary definition – none of those we have considered is. Complex exponentiation is forced upon us by considerations elsewhere, but complex exponentiation along with the trigonometric functions (extended to the complex domain) provide insights into much of the neighbouring mathematics. It does not seem too much of a stretch to suggest we have explanations of some of the central mathematical ideas here (such as an explanation of what exponentiation is really about).

The other kind of generalisation I mentioned is the kind that involves abstracting away from detail to lay bare the crucial features of a mathematical system in question. Perhaps the best-known example here is that of group theory.

Definition 1 (Group). *A group is a set, G, together with an operation $\cdot$ that takes any a and b in G as its arguments and maps them to another element, $a \cdot b$. This set and operation must satisfy the group axioms:*

1. *For all $a, b \in G$, $a \cdot b$ is also in G (Closure Axiom).*
2. *For all $a, b, c \in G$, $(a \cdot b) \cdot c = a \cdot (b \cdot c)$ (Associativity Axiom).*

3. *There exists an $e \in G$, such that for every $a \in G$, the $e \cdot a = a \cdot e = a$ (Identity-Element Axiom).*

4. *For all $a \in G$, there exists $b \in G$ such that $a \cdot b = b \cdot a = e$ (Inverse-Element Axiom).*

One example of a group is the integers under the operation +. The identity element e is 0. Note that the group axioms only capture specific features of the integers. The group axioms ignore other operations such as multiplication and they do not require symmetry of the group operation; $a \cdot b$ is not, in general, the same as $b \cdot a$, but addition on the integers is symmetric. There are also other features of the integers that are ignored (or abstracted away from). For example, integer multiplication is left out of the picture. Other examples of groups are certain spatial rotations and card shuffles. What the axioms allow one to do is study what is common to all such mathematical systems in purely structural terms – the subject matter is not important. It is clear that we learn a great deal about such structures by studying them in this group-theoretic context. For a start, we see connections between the integers and spatial rotations that are not otherwise apparent. A case can be made that such insights can be explanatory.[15]

Both these cases of generalisation – domain extensions and abstraction – look like good testing grounds for the unification account of explanation. They both look to involve precisely the kind of unification we would expect to deliver explanations. But we cannot rely on our philosophical theory to give us the answers about whether we have explanations here; we must look at mathematical practice to see whether our philosophical theory of explanation does a good job of accounting for the practice. Clearly a great deal more work needs to be done. We need to get a clear idea of which proofs are explanatory and which are not, and why. We need to know whether explanations reside in domain extensions and in abstractions. If they do, we need to see if the unification account of explanation does, indeed, do a good job of systematising the mathematical practice in question. Alternative accounts of explanation (e.g., in terms of relevance) also need to be

---

[15] This is similar to what we do in formal logic. Here we are interested in the structure of arguments, not what the arguments are about. We find that we can say something interesting about specific cases from this very abstract structural perspective. We can say that a particular argument is a good one, for instance, because it is an instance of a valid form.

explored. If that's not already enough to keep philosophers of mathematics busy for quite some time, we also need to consider the question of whether mathematics can explain physical phenomena. We turn to this issue in the next section.

5.3 Extra-mathematical explanation

So far we have been discussing intra-mathematical explanation. That is, we have been looking at how one mathematical fact might explain some other mathematical fact, or how mathematical proofs might explain (or fail to explain, as the case may be) the theorems they establish. But now we turn our attention to cases where mathematics might be thought to be explaining physical phenomena – extra-mathematical explanations. To see how this might work, consider an explanatory proof of some mathematical theorem. If that theorem has some physical application, then the proof of the theorem might well explain what's going on in the physical situation. Intra-mathematical explanations can thus 'spill over' into physical applications and become extra-mathematical explanations. For example, consider the Mean-Value Theorem of elementary calculus.

Theorem 5 (Mean Value). *If $f(x)$ is a real-valued function, continuous on the closed interval $[a, b]$ and differentiable on the open interval (a, b), then there is a point $c \in (a, b)$ such that*

$$f'(c) = \frac{f(b) - f(a)}{b - a}.$$

Intuitively, this says that the average rate of change of a continuous function over some interval is instantiated at some point in the interval. There are many physical applications of this theorem. For instance, it guarantees that a plane flying from Sydney to LA with an average speed of 850 km/h (including taxiing, take-off, and landing) at some point in the journey actually has a flying speed of 850 km/h. This theorem, thus, explains why at some point the plane will be flying at its average speed.[16]

Another more scientifically interesting example comes from ecology and involves the life cycles of particular species of cicadas. Philosopher Alan

---

[16] See section 9.2.1 for another similar example of a mathematical theorem invoked to explain a physical fact.

Baker (2005) considers why a particular species of North American cicadas have life cycles which are prime numbers: 13 and 17 years. The explanation of this surprising ecological fact is (arguably) provided by number theory: having a prime-number life cycle is a good strategy for avoiding predators. With a sufficiently large prime cycle, any predators with similar life cycles will very rarely coincide with the most vulnerable stage of the cicada life cycle. It is also interesting to note that the two known cases of this phenomenon yield consecutive prime numbers – 13 and 17 – as the life cycles in question. This suggests that larger primes such as 19, 23, and so on, are impractical for biological reasons. And the smaller primes of 5, 7, and 11 leave the cicadas open to predators with life cycles of 10 years (as well as to predators with life cycles of 15 and 20 years), 14 years, and 22 years respectively. Again it looks as though we have a mathematical explanation on our hands.

In cases like these, it seems that mathematics is carrying the bulk of the explanatory load. Of course we need to have in place interpretations of the mathematics in question, but the explanation of the prime cycles does seem to be mathematical. Examples such as these are not anomalous, either. Once you get a feel for this style of explanation, you find it frequently in science. Let's consider a few more examples.

5.3.1 Honeycomb, asteroids, and contractions

Honeycomb Consider the question of why hive-bee honeycomb has a hexagonal structure. The answer, it turns out, is because of the Honeycomb Theorem: a hexagonal grid represents the most efficient way to divide a surface into regions of equal area with the least total perimeter of cells. (This ancient conjecture – the Honeycomb Conjecture – was first proven and thus promoted to the rank of theorem by the American mathematician Thomas Hales (1958–) in 1999.) There are some biological and pragmatic assumptions required for this explanation to succeed. These include the assumption that bees have a limited supply of wax and need to conserve it while maximising honey storage space. They also need to do this while still being able gain access to the hive from the outside. But with these assumptions in place, the important part of the explanation seems to be purely mathematical and is provided by the Honeycomb Theorem. Any purely biological explanation will be too specific – the latter will be stuck telling the story of how one particular group of bees built one particular hive with a hexagonal structure – and will miss the general point that all hives built under such constraints *must* have a hexagonal structure. The hexagonal structure is a solution to a biological optimisation problem and as such is not a mere accident of any particular hive construction.

The asteroid belt   The Kirkwood gaps are localised regions in the main asteroid belt between Mars and Jupiter where there are relatively few asteroids. The gaps were first noticed by the US astronomer Daniel Kirkwood (1814–95) in 1857. The explanation for the existence and location of these gaps is mathematical and is in terms of the eigenvalues of the local region of the solar system (including Jupiter). The basic idea is that the system has certain resonances and as a consequence some orbits are unstable. Any object initially heading into such an orbit will be dragged off to an orbit on one side or other of its initial orbit as a result of regular close encounters with other bodies (most notably Jupiter). A mathematical analysis delivers both the existence and location of these unstable orbits and thus explains the gaps in question. It's interesting to note that we can seek out a non-mathematical, causal explanation for why each particular asteroid fails to occupy one of the Kirkwood gaps. Each asteroid, however, will have its own complicated, contingent story about the gravitational forces and collisions that that particular asteroid has experienced. Such causal explanations are thus piecemeal and do not tell the whole story. For example, such explanations do not explain why *no* asteroid can maintain a stable orbit in the Kirkwood gaps. The explanation of this important astronomical fact is provided by the mathematics of eigenvalues (that is, basic functional analysis). We thus have scientific statements involving mathematical entities (the eigenvalues of the system) explaining physical phenomena (the relative absence of asteroids in the Kirkwood gaps).[17]

---

[17] If you are unconvinced by this example, because you think that the absence of asteroids does not count as a physical event, consider the case of the collapse of the Tacoma Narrows Bridge in Washington in 1940. This wind-induced collapse is generally thought to be explained by the natural frequencies of the bridge – again, a mathematical explanation. As a result of this bridge collapse, eigenanalysis now features prominently in modern engineering – especially the engineering of suspension-bridge construction. There are many other examples like this one. For example, explaining why an electric

Lorentz contractions   When a body is in motion relative to some observer, the length of the body, as measured by the observer, is seen to decrease in the direction of the motion. This is known as a Lorentz length contraction. The length of the object as measured by the observer is $L' = L\sqrt{1 - v^2/c^2}$, where L is the length of the object as measured in its rest frame, c is the speed of light in a vacuum, and v is the relative velocity between the object and the observer. It can be seen from this formula that the contraction – the difference between L and L' – will be very small unless v is a sufficiently close to c. For example, when $v = c/2$, $L' = \sqrt{3}c/2$, $L' = L/2$ when $v = \sqrt{3}c/2$, and L' approaches zero as v approaches c. The Lorentz contraction is named after the Dutch physicist Hendrik A. Lorentz (1853–1928).

Such Lorentz contractions have been observed and we might reasonably ask for an explanation of them. The explanation comes from special relativity and, in particular, from the Minkowski metric. This is the metric on the space-time manifold of special relativity. The familiar three-dimensional Euclidian metric, which measures the distance between two points, (x_1, y_1, z_1) and (x_2, y_2, z_2) in Euclidean 3-space, is given by $d = \sqrt{(x_1 - x_2)^2 + (y_1 - y_2)^2 + (z_1 - z_2)^2}$. But the Minkowski metric is not simply the four-dimensional analogue of this. When we consider the distance between two space-time points (x_1, y_1, z_1, t_1) and (x_2, y_2, z_2, t_2), where the spatial coordinates are as before and t is the temporal coordinate, a minus sign appears in the metric before the temporal component: $d = \sqrt{(x_1 - x_2)^2 + (y_1 - y_2)^2 + (z_1 - z_2)^2 - (t_1 - t_2)^2}$.[18] This metric is named after the German mathematician Hermann Minkowski (1864–1909), as is the four-dimensional space-time manifold of special relativity: *Minkowski space-time*, or simply *Minkowski space*.

The key to the explanation of the Lorentz length contraction is that the quantity that is preserved in special relativity is the Minkowski length, not the three-dimensional length. Just as three-dimensional objects change their two-dimensional length (their apparent length in your visual field,

guitar feeds back when connected to and placed in front of an amplifier (at a suitably loud volume setting, of course) requires an appeal to the natural resonances of the strings of the guitar, which in turn involves mathematical eigenanalysis. Other examples involve the gaps in the rings of Saturn and shattering wineglasses with loud, high-pitched sounds.

[18] Alternatively, we can keep the plus signs in the metric itself but measure time in purely imaginary numbers – multiples of $i = \sqrt{-1}$.

for instance) when rotated in 3-space, the three-dimensional length of a four-dimensional object in Minkowski space can change. Indeed, having a relative velocity just is a rotation in Minkowski space, so the analogy here is quite strong. In the end, the explanation of the Lorentz contraction is mathematical – it's a geometric explanation involving the structure of Minkowski space and, especially, the Minkowski metric. Such a mathematical explanation should be contrasted with a mechanical explanation which would explain the contraction in terms of forces compressing the body in question. Accepted wisdom has it that the geometric explanation offered by Minkowski space-time is correct and that mechanical explanations are incorrect (or, at least, unnecessary).

5.3.2 What is at issue?

There are a number of interesting issues raised by these examples of mathematical explanations of physical phenomena. The first is the one we noted at the end of section 5.2.2 about the relationship between intra- and extra-mathematical explanation. We might be tempted to treat these as different kinds of explanation but the two are closely connected. What we find is that there can be a kind of 'spillover', where an intra-mathematical explanation can be adapted to play a central role in an extra-mathematical explanation. All we require is that there is an application of the piece of mathematics in question, then explanations from within mathematics flow on to applications. This strongly suggests the need for a unified account of mathematical explanation, in both mathematical and non-mathematical contexts. For example, the explanation for the Lorentz contraction in an abstract Minkowski metric space is surely that geometric explanation offered above. But then it would be very odd indeed to treat the application of the Minkowski metric space to space-time in special relativity differently. If the details of the metric are explanatory in a purely mathematical context, it is hard to see why merely finding an application of the mathematics in question should undermine such explanations. It does look as though we need an account of mathematical explanation that has the potential to function smoothly across both intra- and extra-mathematical contexts.

A related issue arising from extra-mathematical explanations is their relationship to other forms of scientific explanation. Sometimes the quest for a scientific explanation does invite a causal history – what events caused the event we wish to understand. But extra-mathematical explanations do not sit comfortably with such patterns of explanation. Mathematical explanations do not seem to involve causation at all. What are we to say about scientific explanation, as a general category of explanation? Perhaps there is no unified category deserving of the title *scientific explanation*. Perhaps scientific explanation is a hodgepodge of different kinds of explanation – mathematical, causal, and perhaps others. This would suggest a more pluralist attitude towards explanation. Alternatively we might try to develop a unified philosophical account of scientific explanation that includes both causal explanation and mathematical explanation as special cases. Finally, we might try to argue that extra-mathematical explanations don't exist, or that they are reducible in some sense to causal explanation. All of these are live options, but some are more alive than others. (Personally, I don't hold out much hope for the latter, for example.) In any case, it's clear that any satisfying account of scientific explanation will need to address mathematical explanations.

In light of these last issues, we see how mathematical explanations are supposed to lend support to explanatory versions of the indispensability argument for mathematical realism. If mathematics is playing an explanatory role in our scientific theorising, there is a good case to be made for treating the mathematics realistically. After all, one can hardly appeal to electrons in order to explain how an electromagnet works but then deny that electrons exist – not if you're a scientific realist, at least. Similarly, if mathematics is explaining physical phenomena, it would seem that anti-realism about mathematics is untenable. So we find various anti-realists about mathematics trying to defuse the situation by arguing that mathematics is not genuinely explanatory in the way I've been suggesting, or at least that any alleged mathematical explanations can be recast as physical explanations. On the other hand, I and at least some of my mathematical realist colleagues are happy to embrace mathematical explanations of physical phenomena. Either way, mathematical explanations (or the alleged mathematical explanations, if you prefer) have helped breathe life into the debate over the indispensability argument for mathematical realism.

Discussion questions

1. Convert the *reductio* proof of the infinitude of the primes into a direct proof. (Recall that I suggested that some, but not all, *reductio* proofs can be recast as direct proofs and that this was one that could be recast.)
2. Do you think that proofs by mathematical induction are explanatory? Why?
3. Can you give a good, clear example of a mathematical proof that is uncontroversially explanatory? Can you articulate why it is explanatory?
4. Relevance considerations seem to require that explanations stay on topic, while unification invites wider and diverse connections. These two proposals seem to be pulling in different directions. Do you think that they are incompatible accounts of mathematical explanation?
5. Explanations of physical facts are closely related to counterfactuals: A explains B in so far as were A not to have obtained, B would not have obtained. Is there an analogue of this in the mathematical case? Think about what the counterfactuals in question would involve if mathematical facts are necessary truths. Can we make sense of counterfactuals such as: 'if π were rational, then we would be able to square the circle'? (Those who have done some modal logic, think about what the Stalnaker or Lewis semantics for such counterfactuals would look like.)
6. Verify that having a life cycle of 13 or 17 years minimises overlap with predators. What additional assumptions are required?
7. Do you find the (alleged) examples of mathematical explanations of physical phenomena presented in this chapter convincing? Why or why not? Can you think of more convincing examples?
8. If mathematics is genuinely explanatory, precisely how is that supposed to bolster the case for mathematical realism?

Recommended further reading

Below are listed some good places to start for further reading on mathematical explanation. Smart (1990) offers a good introduction to the topic of explanation, introducing the possibility of non-causal explanation. Mancosu (2008c) offers a good overview of mathematical explanation. Some, such as Baker (2005) and Colyvan (2001; 2002), look at mathematical explanation in the context of the indispensability argument. Others, such as Resnik and

Kushner (1987) and Steiner (1978a; 1978b), focus more on mathematical explanation for its own sake.

Baker, A. 2005. 'Are There Genuine Mathematical Explanations of Physical Phenomena?', *Mind*, 114: 223–38.

2009a. 'Mathematical Explanation in Science', *British Journal for the Philosophy of Science*, 60: 611–33.

Batterman, R. W. 2010. 'On the Explanatory Role of Mathematics in Empirical Science', *British Journal for the Philosophy of Science*, 61(1): 1–25.

Colyvan, M. 2001. *The Indispensability of Mathematics*, New York: Oxford University Press, Chapter 3.

2002. 'Mathematics and Aesthetic Considerations in Science', *Mind*, 111: 69–74.

Hafner, J. and Mancosu, P. 2005. 'The Varieties of Mathematical Explanation', in P. Mancosu, K. F. Jørgensen, and S. A. Pedersen (eds.), *Visualization, Explanation and Reasoning Styles in Mathematics*, Dordrecht: Springer, pp. 215–50.

Mancosu, P. 2008b. 'Mathematical Explanation: Why It Matters', in P. Mancosu (ed.), *The Philosophy of Mathematical Practice*, Oxford University Press, pp. 134–49.

2008c. 'Explanation in Mathematics', in E. N. Zalta (ed.), *The Stanford Encyclopedia of Philosophy* (Fall 2008 edn), http://plato.stanford.edu/archives/fall2008/entries/mathematics-explanation/.

Resnik, M. D. and Kushner, D. 1987. 'Explanation, Independence, and Realism in Mathematics', *British Journal for the Philosophy of Science*, 38: 141–58.

Smart, J. J. C. 1990. 'Explanation – Opening Address', in D. Knowles (ed.), *Explanation and Its Limits*, Cambridge University Press, pp. 1–19.

Steiner, M. 1978a. 'Mathematical Explanation', *Philosophical Studies*, 34: 135–51.

1978b. 'Mathematics, Explanation, and Scientific Knowledge', *Noûs*, 12: 17–28.

6   The applicability of mathematics

It is applicability alone which elevates arithmetic from a game to the rank of a science.

Gottlob Frege (1848–1925)[1]

The applications of mathematics in the various branches of science give rise to an interesting philosophical problem: why should physical scientists find that they are unable to even state their theories without the resources of abstract mathematical theories? Moreover, the formulation of physical theories in the language of mathematics often leads to new physical predictions, which were quite unexpected on purely physical grounds. How can turning to the abstract representations of mathematics – far from physical applications – so often turn out to be just what is required in representing and understanding physical systems? In this chapter we will consider the problem of the applicability of mathematics and discuss the prospects for a solution.

6.1 The unreasonable effectiveness of mathematics

The Nobel-Prize-winning Hungarian physicist Eugene Wigner (1902–95) remarked in a famous paper that

> [t]he miracle of the appropriateness of the language of mathematics for the formulation of the laws of physics is a wonderful gift which we neither understand nor deserve. (Wigner 1960, p. 14)

---

[1] G. Frege, 'Gnendgesetze der Arithmetik. Begriffsschriftlich Abgeleitet', in Frege, *Translations from the Philosophical Writings of Gottlob Frege*, ed. P. Geach and M. Black, Cambridge, MA: Blackwell, 1970 [1903], p. 187. Reproduced with kind permission of Blackwell Publishers.

This much-cited passage, however, does not make it clear exactly what the problem is. After all, in order for us to do science we need some language or other. Why shouldn't mathematics be that language? But what is worrying Wigner is something deeper than this. He is concerned about a mismatch between the methodologies of mathematics and physics – physics is driven by empirical evidence and mathematics is in some sense disconnected from the world – yet mathematics still turns out to be just what the physicist needs. Many physicists have followed Wigner and wondered about the relationship between physics and mathematics. For example, British-American physicist Freeman Dyson (1923–) emphasises the fact that mathematics is not merely a calculational tool employed by scientists and that, in some sense, physical theories arise out of the mathematics:

> For a physicist mathematics is not just a tool by means of which phenomena can be calculated; it is the main source of concepts and principles by means of which new theories can be created. (Dyson 1964, p. 129)

The German physicist Heinrich Hertz (1857–94) takes this further, suggesting that mathematics delivers more than what the scientist formalises:

> One cannot escape the feeling that these mathematical formulae have an independent existence and intelligence of their own, that they are wiser than we are, wiser even than their discoverers, that we get more out of them than was originally put into them. (Quoted in Dyson 1964, p. 129)

Another Nobel laureate in physics, Steven Weinberg (1933–), also seems to be pushing in this direction when he suggests that the mathematician, whose work is disconnected from the physical realm, often anticipates the mathematics needed in advanced physical theories:

> It is very strange that mathematicians are led by their sense of mathematical beauty to develop formal structures that physicists only later find useful, even where the mathematician had no such goal in mind… Physicists generally find the ability of mathematicians to anticipate the mathematics needed in the theories of physics quite uncanny. It is as if Neil Armstrong in 1969 when he first set foot on the surface of the moon had found in the lunar dust the footsteps of Jules Verne. (Weinberg 1993, p. 125)

Elsewhere Weinberg suggests that '[i]t is positively spooky how the physicist finds the mathematician has been there before him or her' (Weinberg 1986, p. 725).

It is not only physicists who've worried about the relationship between mathematics and physics. Philosopher Mark Steiner also believes that there is a problem worthy of attention here:

> [H]ow does the mathematician – closer to the artist than the explorer – by turning away from nature, arrive at its most appropriate descriptions? (Steiner 1995, p. 154)

Here Steiner suggests that the core problem arises from the different methodologies of mathematics and physics. Before we pursue the problem, thus construed, let's get clearer about what the target is by setting aside a couple of other puzzles involving the applications of mathematics.

Mark Steiner rightly points out that Wigner's 'puzzle' is in fact a whole family of puzzles, and these are not clearly distinguished by Wigner; it all depends on what you mean by 'applicability' when talking of the applications of mathematics. It is important to distinguish the different senses of 'applicability', because some of the associated puzzles are easily solved while others are not. For example, there is the problem of the (semantic) applicability of mathematical theorems – the problem of explaining the validity of mathematical reasoning in both pure and applied contexts.[2] Indeed, for every way that mathematics is applicable to science, we potentially have a Wigner-style problem. It might be an interesting exercise to try to articulate all the ways in which mathematics is applicable to science and see which of these leads to a Wigner-style puzzle. For present purposes, however, we can set this task aside and cut straight to one of the more serious applicability problems – one gestured towards in several of the quotations above.

The core problem, it seems to me, is that of explaining the appropriateness of mathematical concepts – concepts developed by a-priori methods – for the description of the physical world. Of particular interest here are cases where the mathematics seems to be playing a crucial role in making predictions. Mark Steiner has argued for his own version of Wigner's

---

[2] To explain, for instance, why the truth of (i) there are 11 Lennon–McCartney songs on the Beatles' 1966 album *Revolver*, (ii) there are 3 non-Lennon–McCartney songs on that same album, and (iii) $11 + 3 = 14$, implies that there are 14 songs on Revolver. (The problem is that in (i) and (ii) '11' and '3' seem to act as names of predicates and yet in (iii) '11' and '3' apparently act as names of objects. What we require is a constant interpretation of the mathematical vocabulary across such contexts.)

thesis along these lines. According to Steiner, the puzzle is not simply the extraordinary appropriateness of mathematics for the formulation of physical theories, but concerns the role mathematics plays in the very discovery of those theories. In particular, we require an explanation of this puzzle that is in keeping with the methodology of mathematics – a methodology that does not seem to be guided at every turn by the needs of physics.

Thus construed, the problem is epistemic: why is mathematics, which is developed primarily with broadly aesthetic considerations in mind, so crucial in both the discovery and the statement of our best physical theories? Put thus, the problem may seem like one manifestation of a more general problem in the philosophy of science, namely the problem of justifying the appeal to aesthetic considerations, such as simplicity, elegance, and so on.

Scientists and philosophers of science invoke aesthetic considerations to help decide between two theories. For example, when two theories are equally good in other respects, we generally prefer the simpler theory. But aesthetics play a much more puzzling role in the Wigner/Steiner problem. Here aesthetic considerations are largely responsible for the development of mathematical theories. These, in turn (as we shall see shortly), play a crucial role in the discovery of our best scientific theories. In particular, novel *empirical* phenomena are discovered via mathematical analogy. In short, aesthetic considerations are not just being invoked to decide between two theories; they seem to be an integral part of the process of scientific discovery. The role of aesthetics in the Wigner problem is quite different from whatever puzzles there may be about the role of aesthetics in scientific theory choice.

An example will help. James Clerk Maxwell (1831–79) found that the accepted laws for electromagnetic phenomena prior to about 1864, namely Gauss's law for electricity, Gauss's law for magnetism, Faraday's law, and Ampère's law, jointly contravened the conservation of electric charge. Maxwell thus modified Ampère's law to include a *displacement current*, which was not an electric current in the usual sense (a so-called *conduction current*), but a rate of change (with respect to time) of an electric field. This modification was made on the basis of formal mathematical analogy, *not* on the basis of empirical evidence. Indeed, there was very little (if any) empirical evidence at the time for the displacement current. The analogy was with Newtonian gravitational theory's conservation of mass principle.

The modified Ampère law states that the curl[3] of a magnetic field is proportional to the sum of the conduction current and the displacement current:

$$\nabla \times \mathbf{B} = \frac{4\pi}{c}\mathbf{J} + \frac{1}{c}\frac{\partial}{\partial t}\mathbf{E}. \tag{6.1}$$

Here $\mathbf{E}$ and $\mathbf{B}$ are the electric and magnetic field vectors respectively, $\mathbf{J}$ is the current density, and c is the speed of light in a vacuum.[4] When this law (known as the Maxwell–Ampère law) replaces the original Ampère law in the above set of equations, they are known as *Maxwell's equations* and they bring an elegant unity to the subject of electromagnetism.

The interesting part of this story for the purposes of the present discussion, though, is that Maxwell's equations were formulated on the assumption that the charges in question moved with a constant velocity, and yet such was Maxwell's faith in the equations that he assumed that they would hold for any arbitrary system of electric fields, currents, and magnetic fields. In particular, he assumed that they would hold for charges with accelerated motion and for systems with zero conduction current. An unexpected consequence of Maxwell's equations followed in this more general setting: a changing magnetic field would produce a changing electric field and vice versa. Again from the equations, Maxwell found that the result of the interactions between these changing fields on one another is a wave of electric and magnetic fields that can propagate through a vacuum. He thus predicted the phenomenon of electromagnetic radiation. Furthermore, he showed that the speed of propagation of this radiation is the speed of light. This was the first evidence that light was an electromagnetic phenomenon.

These predictions, confirmed experimentally by Heinrich Hertz in 1888, can be largely attributed to the mathematics. The predictions in question were being made for circumstances beyond the assumptions of the equations' formulation. Moreover, the formulation of the crucial equation (the Maxwell–Ampère law) for these predictions was based on formal mathematical analogy. Cases such as this seem puzzling, at least when presented

---

[3] The curl, ∇, is a differential operator on vectors. If $\mathbf{F} = \left[F_x, F_y, F_z\right]$ is a vector in a three-dimensional Euclidean space, with unit vectors $\mathbf{i}$, $\mathbf{j}$, and $\mathbf{k}$, $\nabla \times \mathbf{F} = (\partial F_z/\partial y - \partial F_y/\partial z)\mathbf{i} + (\partial F_x/\partial z - \partial F_z/\partial x)\mathbf{j} + (\partial F_y/\partial x - \partial F_x/\partial y)\mathbf{k}$.

[4] The first term on the right of equation 6.1 is the conduction current and the second on the right is the displacement current.

a certain way. The question is whether the puzzlement here is merely an artefact of the presentation (perhaps because some particular philosophy of mathematics is explicitly or implicitly invoked).

It is interesting to note that in Wigner's article he seems to be taking a distinctly anti-realist attitude towards mathematics. His attempted definition of 'mathematics' illustrates this:

> [M]athematics is the science of skillful operations with concepts and rules *invented* just for that purpose. (Wigner 1960, p. 2; my italics)

Others, such as Reuben Hersh, also adopt anti-realist language when stating the problem:[5]

> There is no way to deny the obvious fact that arithmetic was *invented* without any special regard for science, including physics; and that it turned out (unexpectedly) to be needed by every physicist. (Hersh 1990, p. 67; my italics)

Some, such as Paul Davies (1992, pp. 140–60) and Roger Penrose (1989, pp. 556–7), have suggested that the unreasonable effectiveness of mathematics in the physical sciences is evidence for realism about mathematics. That is, there is only a puzzle here if you think we invent mathematics and then find that this invention is needed to describe the physical world.

Given such comments we might well think that Wigner's puzzle brings us back to the realism–anti-realism debate. For what it's worth, I think the Wigner puzzle cuts across the realist–anti-realist divide in the philosophy of mathematics. After all, as I suggested earlier, the central problem is that mathematics seems to proceed via a-priori means and yet finds applications in a-posteriori science. Thus stated, the problem has little, if anything, to do with realism and anti-realism issues. It is an epistemic problem or, if you prefer, a problem about the different methodologies of mathematics and empirical science. In any case, both realists and anti-realists need to provide an account of applied mathematics. We can thus set the thorny realism–anti-realism issues

---

[5] Recall also Weinberg's reference to Jules Verne in the passage I quoted earlier and Steiner's remark (again quoted earlier) about the mathematician being more like an artist than an explorer.

aside and focus on coming to a better understanding of mathematics in applications.

6.2 Towards a philosophy of applied mathematics

In this section we consider some responses to the Wigner problem and sketch an account of applied mathematics that might make the effectiveness of mathematics seem less unreasonable.

6.2.1 Why so unreasonable?

One line of response to Wigner is to deny that mathematics is all that effective in physical science. Sure, mathematics is (sometimes) effective, but is it unreasonably so? There are various ways to deny the unreasonableness claim. One way is to stress the difficulties involved in getting a mathematical model to work. Anyone who has done any mathematical modelling knows how difficult it is to get the mathematics and the world to see eye to eye. The unreasonable effectiveness problem thus arises from overlooking such difficulties and looking at the end product – the fully developed mathematical models and mathematised scientific theories – and forgetting, or not appreciating, the painstaking work that went into developing the theories and models in question. Such work often involves developing new mathematics, deliberately ignoring particularly uncooperative bits of the world, and occasionally riding roughshod over some of the mathematics. Once we look at the details, so the suggestion goes, we will be less impressed by how effective mathematics is.

Another, related response is to suggest that we do not have a very good grip on what would count as unreasonable. After all, if someone is unreasonably effective at predicting the outcomes of fair coin tosses, say, they would be expected to correctly predict the outcome more often than not. It won't do to simply note that they get it right sometimes and thus conclude that they are unreasonably effective at predicting coin tosses. We need to do some statistics before we can determine whether there is anything unreasonable here. So too with the Wigner puzzle. All Wigner or anyone else has done is draw attention to some admittedly stunning success stories. But what about all the failures? We need to determine what would count as an unreasonable success rate and see if the statistics for the historical cases in science bear out the unreasonableness. There are, of course, many obstacles to such an investigation. For a start, typically only the success stories get published. A paper pointing out that an attempt to model some physical phenomenon with a particular piece of mathematics was a dismal failure is not likely to be published. But surely there are many such attempts preceding the success stories we know about? So there is a selection bias here: a non-representative sample that is in effect selecting itself. The failures are invisible. To get the failures we need access to the waste-paper baskets of scientists, past and present. In any case, until we know more about the relationship of successes to failures, we simply do not know whether the success of mathematics is unreasonable or not.[6]

Another response along these lines is to suggest that, in some sense, you get what you look for. The best tools we have for modelling the world are mathematical tools, such as differential equations, so we tend to shoehorn the world into this framework. Put differently we tackle only the physical problems that are amenable to the mathematical methods we have at our disposal. To the Pythagoreans, the world looked as though it was all natural numbers; these days, it looks as though it's all complex Hilbert spaces and differential equations. The English physicist Arthur Eddington (1882–1944) once wrote:

> Let us suppose that an ichthyologist is exploring the life of the ocean. He casts a net into the water and brings up a fishy assortment. Surveying his catch, he proceeds in the usual manner of a scientist to systematise what it reveals. He arrives at two generalisations:
>
> (1) No sea-creature is less than two inches long.
> (2) All sea-creatures have gills.
>
> These are both true of his catch, and he assumes tentatively that they will remain true however often he repeats it. (Eddington 1939, p. 16)

Although Eddington goes on to make a different point in relation to the analogy, it stands as a very nice illustration of how our observational and data collection methods can influence what we find. The relevance to the

---

[6] There's an old joke about a physicist approaching a funding agency for a multi-million-dollar piece of equipment. The director of the funding agency shakes her head and says: 'You physicists are always after money; why can't you be more like the mathematicians? All they need are pencils, paper, and waste-paper baskets. Better still, why can't you be like the philosophers? They just need pencils and paper.'

present point is that if we look at the world through mathematical lenses we tend to see a mathematical world. Or, at least, the bits of the world that are amenable to mathematisation are easier to see when wearing such lenses.[7] Perhaps Wigner is making such a mistake. Just like Eddington's ichthyologist with only a specific net available to him to investigate the contents of the ocean, Wigner has only one tool available to him in the investigation of the world – mathematics. And like the ichthyologist, Wigner misses interesting phenomena because of the limitations of his investigative tools. But the limitations should be recognised and overgeneralisations based on them avoided. Famously, the psychologist Abraham Maslow (1908–70) expressed the point this way: 'It is tempting, if the only tool you have is a hammer, to treat everything as if it were a nail' (Maslow 1966, p. 15).

These few brief and rather sketchy responses to Wigner's puzzle are at best only partial solutions.[8] They mostly consist in suggesting that the alleged unreasonableness is not all that unreasonable after all, or at least, may not be so unreasonable once seen in the full context. That is, they are attempts to defuse the puzzle. But perhaps it would be better to tackle the issue of the applicability of mathematics head-on. Irrespective of how unreasonably effective mathematics is in the physical sciences, it would be good to have a better philosophical understanding of mathematics in applications. In the next section I consider one attempt to develop an account of applied mathematics – the mapping account. This account is in its early stages of development, but it is worth considering, if only to help us get a clearer picture of what the relevant issues are.

6.2.2 Maps and mathematics

A satisfactory philosophical account of applied mathematics should answer all questions about the application process. For one thing, it should resolve the puzzle raised by Wigner, but perhaps we shouldn't make the Wigner puzzle the primary target. If we set out merely to answer Wigner, we may end up with a philosophy of applied mathematics that's tailor-made for

---

[7] This is an example of theory-laden observation, where what you look for and the interpretation of what you see is determined to some extent by the background theory.

[8] They are mostly derived from Hamming (1980), where other tentative suggestions about how to answer Wigner can be found. See also Azzouni (2000) and Wilson (2000).

one and only one task. A better strategy might be to come to a better understanding of applied mathematics generally – including an appreciation of the range of issues involved – and try to advance a positive account that answers Wigner's and other challenges in passing.

So what would a philosophical account of mathematics in applications look like? One popular starting point is to think of mathematical models in the broader context of scientific and other models. Models typically preserve certain structural features of the target system. So, for example, the applicability of real analysis to flat space-time is explained by the structural similarities between $\mathbb{R}^4$ (with the Minkowski metric) and flat space-time. This is similar to the way a map of London is useful because it preserves some salient features of London, such as scaled distances and directions. Notice that maps do not need to preserve all features of a city. For instance, city maps do not usually represent every building and certainly do not represent people's locations. Some maps even give up on faithfully representing spatial relations. The standard London Underground map is like this.

Thinking of a mathematical model in these terms can be useful. The mathematical model needs to represent some aspects of the structure of the target system but need not, and probably should not, faithfully represent all aspects of the target system. This is an obvious enough point, but some have insisted that mathematical models work because they are isomorphic to the target system. This is clearly a mistake. Models of any kind, whether they be maps of cities or mathematical models of a damped pendulum, can be useful without being isomorphic to the target system. Sometimes the target system has more structure that's not represented in the model and sometimes the mathematics has more structure. The former is familiar enough, but the latter requires some elaboration.

Models are not always simplifications in the sense that they preserve only some of the structure of the target system. For example, when modelling a fluid flowing through a pipe, standard models, familiar from undergraduate mathematics, employ differential equations, (such as the Navier–Stokes equations), which treat the fluid as a continuous medium. But fluids consist of discrete molecules, so the mathematical model has a much richer structure than the target system. Oddly this extra structure is there for reasons of simplicity – the relevant differential-equation model is easier to work with than a corresponding discrete difference-equations model – but the simplicity in question is clearly not the simplicity of less structure. Quite the opposite; it's the simplicity afforded by a richer structure.

Often mathematical models will have richer structure in one part but less structure in another. Again, fluid-mechanics models serve us well here. We've already seen how the model can have more structure than the target system (the former continuous while the latter is discrete), but the model can simultaneously have less structure. For instance, climate models are big fluid-mechanics models and they ignore many features of the target system, such as local eddy currents, people sneezing, and butterflies flapping their wings. This makes it very difficult to spell out, in any systematic way, what the nature of the structural similarity between model and target system is. It certainly isn't isomorphism, and even weaker notions such as homomorphism, epimorphism, and monomorphisms are problematic.[9]

Apart from anything else, mathematics is a rich source of structures – $\mathbb{R}^n$, the ZFC hierarchy, Hausdorff topological spaces, and so on. These structures can be taken from the shelf, so to speak, in order to model various target systems, or serve as the starting point of a model of some target system. Mathematics is thus like a very rich map repository, which can be called upon when a new city map is required. Either the map repository will have the map of the city in question or it will have a map that's close enough and which can be modified to do the job. This view of applied mathematics has mathematics serving as a source of structures for scientific theorising, with mappings (in the mathematical sense) between the mathematical model and the target systems. These mappings ensure that crucial structural features of the target system are mirrored in the mathematical model. This has become known as *the mapping account of applied mathematics*.[10]

---

[9] A *homomorphism* is a mapping φ from one structure A to another structure B, and φ preserves some class of structural relations on A. That is, a homomorphism maps not only the objects of one domain to another, it does so in a way that preserves certain structural relations between the objects in question. An *epimorphism* is a surjective homomorphism (i.e., every member of B is the image under φ of at least one member of A); a *monomorphism* is a injective homomorphism (i.e., different members of A are mapped to different members of B).

[10] See Leng (2002; 2010) and Pincock (2004; 2007) for defences of such an account, and Batterman (2010) and Bueno and Colyvan (2011) for criticism of the mapping account and for alternative accounts.

There is something undoubtedly right about the mapping account; it seems right as far as it goes, but it does not (yet) go very far. There needs to be more work on the nature of the mappings in question. For example, we need to know what kind of structure-preserving mappings are typically involved between the mathematical model and the target system. When the mathematical model has more structure than the target system, how do we go about interpreting the extra structure in question? Should it always be dismissed as mere artefact of the model? This does not seem right; Wigner's problem alerted us to the possibility of novel predictions coming from the mathematics. Sometimes these predictions come from initially uninterpreted parts of the model and have no physical precedence (e.g., Maxwell's prediction of electromagnetic waves propagated through a vacuum). How is it that adding structure can simplify the model (as in using differential equations to model discrete phenomena)? And how can abstracting away from causal detail, as so often happens in mathematical models, advance our understanding of concrete physical situations? These are all questions that need answers. Thus far, no one has an account of applied mathematics that can provide satisfactory answers to all these questions.

6.3 What's maths got to do with it?

Let's not forget that physics is not the only consumer of mathematics. In many of the special sciences, mathematical models are used to provide information about specified target systems. For instance, population models are used in ecology to make predictions about the abundance of real populations of particular organisms. The status of mathematical models in ecology, though, is unclear and their use is hotly contested by some practitioners. A common objection levelled at the use of these models is that they ignore all the known, causally relevant details of the often complex target systems. Indeed, the objection continues, mathematical models, by their very nature, abstract away from what matters and thus cannot be relied upon to provide any useful information about the systems they are supposed to represent. In this section we examine the role of some typical mathematical models in population ecology. In a sense, these models do ignore the causal details, but this move can not only be justified, it is necessary. I will argue that idealising away from complicating causal details often gives a clearer view of what really matters. And often what really matters is not the push and shove of base-level causal processes, but higher-level predictions and (non-causal) explanations.

6.3.1 Case study: population ecology

Population ecology is the study of population abundance and how this changes over time. For present purposes, a population can be thought of as a collection of individuals of the same species, inhabiting the same region. Population ecology is a high-level special science, but relies heavily on mathematical models. (It is thus a soft science in one sense – in the sense of being high-level and quite removed from physics – but in another sense it is a hard science – in the sense that it is mathematically sophisticated.) There are a number of issues associated with applying mathematics to population ecology, but the focus here will be on an issue that is of significance for working ecologists and has a direct bearing on the way they go about their business.[11]

It will be useful to present a couple of typical mathematical models, of the kind we are interested in. First consider *the logistic equation*. This is a model of a single population's abundance, N–exponential at first and then flattening out as it approaches carrying capacity, K:

$$\frac{dN}{dt} = rN\left(1 - \frac{N}{K}\right),$$

where r, is the population growth rate and t is time.

Another key example is *the Lotka–Volterra model*. This models the population of a predator and its prey via a pair of coupled first-order differential equations:

$$\frac{dV}{dt} = rV - \alpha V P$$
$$\frac{dP}{dt} = \beta V P - q P.$$

Here V is the population of the prey, P is the population of the predator, r is the intrinsic rate of increase in prey population, q is the per capita death rate of the predator population, and α and β are parameters: the capture

---

[11] See, for example, Levins (1966) and May (2004) for some discussion on this and related issues by a couple of very prominent ecologists.

efficiency and the conversion efficiency, respectively. These equations can give rise to complex dynamics, where the dual, out-of-phase population oscillations of predator and prey are the best known.

Of course both these mathematical models are overly simple and are rarely used beyond introductory texts in population ecology. For example, the logistic equation treats the carrying capacity of the environment as constant, and the Lotka–Volterra equations treat the predators as specialists, incapable of eating anything other than the prey in question. Both these assumptions are typically false. These models do, however, serve as the basis for many of the more realistic models used in population ecology. The more sophisticated models add complications such as age structure, variable growth rates, and the like. These complications do not matter for our purposes, though. Even in these more complicated models, biological detail is deliberately omitted, yet the models are adequate for the purposes for which they are intended. The issues we're interested in are easier to see in the simpler models, so let's stick with those.

6.3.2 Birth, death, and mathematics

We are now in a position to give voice to a philosophical problem arising from the use of mathematical models in population ecology. Population abundance is completely determined by biological facts at the organism level – births, deaths, immigration, and emigration – but the (standard) mathematical models leave out all the biological detail of which individuals are dying (and why), which are immigrating (and why), and so on. That is, the mathematical models ignore the only things that matter, namely, the biological facts. The mathematical models here – the relevant differential equations – seem to ignore the biology, and yet it is the biology that fully determines population abundances. How can ignoring that which is most important ever be a good modelling strategy?

We might recast the problem in terms of explanation: the mathematical models ignore the causal detail and thus would seem to lack explanatory power. The model may tell us that the abundance of some population at time t is N, but without knowing anything about the organism-level biology, we will not know *why* the population at time t is N and will have little confidence in such predictions. A full account of the relevant biology, on the other hand, would include all the causal detail and *would* provide the required explanations. Let's focus on this explanatory version of the puzzle, because I think it is what underwrites the less-specific worries expressed in the previous paragraph.

Now consider what makes these mathematical models in ecology tick. There is no reason to suggest that these models are not explanatory. There are three different ways in which the models in question could explain. First, note that the mathematical models do not ignore the biological detail; at least sometimes the models in question are offering biological explanations, albeit explanations couched in mathematical terms. Second, understanding a system often involves ignoring, or rather abstracting away from, causal detail in order to get the right perspective on it. Finally, we might encounter cases of mathematical explanations of empirical phenomena (as discussed in Chapter 5).

Recall that we started out with the charge that mathematical models leave out all the relevant biological detail. But this is not quite right. Often the mathematical model is just representing the biology in a mathematical form. For example, in the logistic equation, all the information about births, deaths, immigration, and emigration is packed into r and all the information about the resources is packed into the constant K. The information about the predators' impact on the per capita growth rate of the prey is summarised in the Lotka–Volterra equation by α – the capture efficiency parameter – and the information about the predators' ability to turn prey into per capita growth of the predator population is summarised by β – the conversion efficiency parameter.

You might have misgivings about the representation of this information. You might, for example, object that r and K are represented in the logistic model as constants. But this is a different objection. This is a concern about the simplicity of the model. As I mentioned earlier, we can provide more complex models that relinquish some of the more unrealistic idealisations. These more complex models also have their idealisations, though. Indeed, it is part of the very enterprise of modelling that some details are ignored. So the basic concern about biological detail not being represented in the mathematical models under consideration is misplaced. Of course not all the biological detail is present in the model, but the fact remains that many of the key terms of the mathematical models have natural biological interpretations, or at least are representing or summarising the biological information in mathematical form. The mathematical models have a lot more biology in them than you might at first think.

In cases where the biology is represented in mathematical form, the model is indeed capable of offering perfectly legitimate biological explanations. For instance, think of the standard story of how population cycles arise as a result of predator–prey interactions. The cycles in question are solutions to the coupled differential equations in question, but there is also a very natural biological explanation that can be extracted from the mathematical model: when the predator population is high the predators catch many of the prey so that the latter's population falls, but then there is less food for the predators, so after a time the predator population also falls; but now there is less pressure on the prey population, so it recovers and this, in turn, supports an increase in the predator population (after a similar time lag). This cyclic behaviour falls out of the mathematics, but the explanation, once suitably interpreted, is in fact a perfectly respectable ecological explanation.

Next, notice that ignoring some detail can lead to insights via analogy.[12] Sometimes similarities between systems will not be apparent until certain details are ignored. Mathematics is particularly well suited to drawing out such similarities, because mathematics allows one – indeed forces one – to abstract away from the causal detail and notice abstract similarities. For example, Newton's law of cooling/heating is just the logistic equation with abundance replaced by temperature of the body in question, and carrying capacity replaced by ambient room temperature. Why are such connections between systems important? One reason is that it saves work: one can import results already at hand from work done elsewhere. Once the connection between the logistic equation and the cooling/heating equation is recognised, for example, results from either area can be used by the other (suitably interpreted, of course). Moreover, these rather abstract connections – often only apparent via the mathematics – can lead to new developments and even help with explanations.

We have already seen that mathematics can be the vehicle for delivering biological explanations, but often the mathematics can facilitate more transparent explanations. Mathematical models can sometimes do more

---

[12] See Colyvan and Ginzburg (2010) for more on analogical reasoning in ecology.

than just represent the biology in mathematical form, then deliver essentially biological explanations of biological facts (albeit in mathematical guise). Sometimes the mathematics delivers explanations that would not be apparent otherwise. For example, the explanation of the different kinds of complex behaviour a population can exhibit as it approaches its carrying capacity – damped oscillations, asymptotic approach, overshooting, and crashes – may be best seen via the mathematics of the logistic equation.

Finally, as we discussed in Chapter 5, there are reasons to think that there can be genuinely mathematical explanations of empirical facts. The examples of the hexagonal structure of honeycomb and the prime life cycles of the cicadas are arguably mathematical explanations of biological facts. Another example of a mathematical explanation in ecology will help and will also illustrate how analogical reasoning can play an important role in delivering the mathematical explanation in question. As we saw earlier, population cycles are one of the better-known solutions of the Lotka–Volterra equations, but there are other, more general models of population cycles. The more general models invoke a second-order differential equation (instead of the coupled first-order equations in the Lotka–Volterra model) and allow for single-species population cycles (Ginzburg and Colyvan 2004). This more general approach to population cycles is mathematically very similar to periodic solutions to two-body problems in celestial mechanics. This interdisciplinary connection is interesting in its own right, but it is much more than a mere curiosity. This analogy has the potential to drive a number of developments in population ecology.

First, the similar mathematical treatment suggests that there ought to be an ecological counterpart of inertia in physics, and this has led to investigations into 'ecological inertia'.[13] A second development arising from the analogy in question is that there should be stable and unstable orbits, as is the case with satellite orbits. In the rings of Saturn, for instance, there are well-defined gaps marking out the unstable orbits of this system. Similarly, in the asteroid belt between Mars and Jupiter there are gaps – the Kirkwood gaps – and these, as we have seen, represent unstable orbits as a result of resonance effects with other massive bodies (most notably Jupiter). One might well expect to see similar gaps in population cycles and these

---

[13] These are essentially cross-generational time lags in population responses to changes in environment (Inchausti and Ginzburg 2009).

gaps, if they exist, would be explained mathematically, by appeal to very general structural features of the systems in question. Not only would such explanations be mathematical, they would have been discovered by way of an analogy, facilitated by the mathematics in question.

If this is right and such cases are indeed cases where mathematics is carrying the bulk of the explanatory load, there is still the question of how mathematics can do this. There are several possibilities: (i) mathematics can demonstrate how something surprising is possible (e.g., stable two-species population cycles); (ii) mathematics can show that under a broad range of conditions, something initially surprising must occur (e.g., hexagonal structure in honeycomb); (iii) mathematics can demonstrate structural constraints on the system, thus delivering impossibility results (e.g., certain population abundance cycles are impossible); (iv) mathematics can demonstrate structural similarities between systems (e.g., missing population periods and the gaps in the rings of Saturn).

In light of the preceding discussion, it is a mistake to assume that because mathematical models ignore some of the biological detail they are not capable of delivering explanations. Indeed, to deliver the explanation in at least some of these cases might *require* that some biological detail be ignored.[14] Given the modal character of the three kinds of explanation just mentioned (involving possibility, necessity, and impossibility), it is hard to see how any causal explanation can deliver such explanations.

Discussion questions

1. Why do you think Wigner and others used anti-realist language in articulating the problem of the unreasonable effectiveness of mathematics? Is the problem more pressing for the anti-realist or for the realist?
2. Is it correct to see the Wigner problem as cutting across the realism–anti-realism debate? Is there any reason to expect a realist account of Wigner's problem to be substantially different from an anti-realist one? Spell out a version of the problem for each.
3. Consider an evolutionary response to the Wigner puzzle, according to which the natural world selects for mathematical ability. According to

[14] See Batterman (2002a; 2002b; 2010) for more on the role of abstraction in such explanations.

this line of thought, it would be no surprise that we have the mathematical abilities we do and that we use these abilities to model the world. Flesh out an argument along these lines. How good is it?

4. Do the details included in, and excluded from, a model depend on the purpose of the model? What purposes can models serve?

5. Make a case for treating extra mathematical structure in models as being mere artefact. Does this simplify the kinds of mappings we need to entertain between the model and the world?

6. Why don't false assumptions (e.g., that populations of animals are continuous) in a model invalidate the model? How can known-to-be-false models be trusted?

7. In Chapter 4 I argued that metaphors can't explain unless they are standing proxy for a real explanation. But in section 6.3.2 I suggested that abstract mathematical models *can* explain, even if they are not merely standing proxy for a deeper biological explanation. Is there a tension here?

Recommended further reading

The classic source for the problem of the unreasonable effectiveness of mathematics is Wigner's (1960) original paper. For a philosopher's perspective on the problem see Steiner (1989, 1995). See also Grattan-Guinness (2008), Hamming (1980), and Wilson (2000) for more on the unreasonable effectiveness of mathematics. See Leng (2002) and Pincock (2004) for more on the mapping account of mathematics. See Batterman (2002b) for more on the role of abstraction in mathematical models. Mathematical models in ecology are discussed in Colyvan and Ginzburg (2010), Levins (1966), and May (2004).

Batterman, R. W. 2002b. 'Asymptotics and the Role of Minimal Models', *British Journal for the Philosophy of Science*, 53(1): 21–38.

Colyvan, M. and Ginzburg, L. R. 2010. 'Analogical Thinking in Ecology: Looking beyond Disciplinary Boundaries', *Quarterly Review of Biology*, 85(2): 171–82.

Grattan-Guinness, I. 2008. 'Solving Wigner's Mystery: The Reasonable (Though Perhaps Limited) Effectiveness of Mathematics in the Natural Sciences', *Mathematical Intelligencer*, 30: 7–17.

Hamming, R. W. 1980. 'The Unreasonable Effectiveness of Mathematics', *American Mathematical Monthly*, 87(2): 81–90.

Leng, M. 2002. 'What's Wrong with Indispensability? (Or the Case for Recreational Mathematics)', *Synthese*, 131: 395–417.

Levins, R. 1966. 'The Strategy of Model Building in Population Biology', *American Scientist*, 54: 421–31.

May, R. M. 2004. 'Uses and Abuses of Mathematics in Biology', *Science*, 303 (6 February): 790–3.

Pincock, C. 2004. 'A New Perspective on the Problem of Applying Mathematics', *Philosophia Mathematica*, 12: 135–61.

Steiner, M. 1989. 'The Application of Mathematics to Natural Science', *Journal of Philosophy*, 86: 449–80.

    1995. 'The Applicabilities of Mathematics', *Philosophia Mathematica*, 3: 129–56.

Wigner, E. P. 1960. 'The Unreasonable Effectiveness of Mathematics in the Natural Sciences', *Communications on Pure and Applied Mathematics*, 13: 1–14.

Wilson, M. 2000. 'The Unreasonable Uncooperativeness of Mathematics in the Natural Sciences', *The Monist*, 83: 296–315.

7 Who's afraid of inconsistent mathematics?

In formal logic, a contradiction is the signal of a defeat: but in the evolution of real knowledge it marks the first step in progress towards a victory.

Alfred North Whitehead (1861–1947)[1]

Contemporary mathematical theories are generally thought to be consistent. But it hasn't always been this way; there have been times when the consistency of mathematics has been called into question. Some theories, such as naïve set theory and (arguably) the early calculus, were shown to be inconsistent. In this chapter we will consider some of the philosophical issues associated with inconsistent mathematical theories.

7.1 Introducing inconsistency

7.1.1 A five-line proof of Fermat's Last Theorem

Fermat's Last Theorem states that there are no positive integers x, y, and z, and integer $n > 2$, such that $x^n + y^n = z^n$. This theorem has a long and illustrious history but was finally proved in the 1990s by English mathematician Andrew Wiles (1953–). Despite the apparent simplicity of the theorem itself, the proof runs to over a hundred pages, invokes some very advanced mathematics (the theory of elliptic curves, among other things), and is understandable to only a handful of mathematicians.[2] But consider the following proof of this theorem.

Theorem 6 (Fermat's Last Theorem (FLT)). *There are no positive integers x, y, and z, and integer $n > 2$, such that $x^n + y^n = z^n$.*

---

[1] A. N. Whitehead, *Science and the Modern World*, The Macmillan Company (and Cambridge University Press), 1925, p. 231.

[2] See Singh (1997) for a popular account of Fermat's Last Theorem.

Proof. Let $\mathcal{R}$ stand for the Russell set, the set of all things that are not members of themselves: $\mathcal{R} = \{x : x \notin x\}$. It is straightforward to show that this set is both a member of itself and not a member of itself: $\mathcal{R} \in \mathcal{R}$ and $\mathcal{R} \notin \mathcal{R}$. So since $\mathcal{R} \in \mathcal{R}$, it follows that $\mathcal{R} \in \mathcal{R}$ or FLT. But since $\mathcal{R} \notin \mathcal{R}$, by disjunctive syllogism, FLT.   □

This proof is short, easily understood by anyone with just a bit of high-school mathematics. Moreover, the proof was available to mathematicians well before Wiles's groundbreaking research. Why wasn't the above proof ever advanced? One reason is that the proof invokes an inconsistent mathematical theory, namely, naïve set theory. This theory was shown to be inconsistent early in the twentieth century. The most famous inconsistency arising in it was Russell's paradox, which invoked the same problematic set as used in the above proof.[3] As we saw earlier, paradoxes such as Russell's (and others such as the Burali-Forti ordinal paradox and Cantor's cardinality paradox) led to a crisis in mathematics at the turn of the twentieth century. This, in turn, led to many years of sustained work on the foundations of mathematics. In particular, a huge effort was put into finding a consistent (or at least not known-to-be-inconsistent) replacement for naïve set theory. The now generally agreed upon replacement is Zermelo–Fraenkel set theory with the axiom of choice (ZFC).[4]

But the inconsistency of naïve set theory cannot be the whole story of why the above proof of Fermat's Last Theorem was never seriously advanced. After all, there was a period of some 30-odd years between the discovery of Russell's paradox and the development of ZFC. Mathematicians did not shut up shop until the foundational questions were settled. They continued working, using naïve set theory, albeit rather cautiously. Moreover, it might be argued that many mathematicians to this day still use naïve set theory, or something very much like it. After all, as long as you are careful to skirt around the known paradoxes, naïve set theory can be safely used in areas such as analysis, topology, algebra, and the like. Indeed, most mathematical proofs, outside of set theory, do not explicitly state the set theory being employed. Moreover, typically they do not show how the various set-theoretic constructions are legitimate according to ZFC. This suggests,

---

[3] Recall that the paradox is that the Russell set both is and is not a member of itself.

[4] See Giaquinto (2002) for an account of the history and Enderton (1977) for details of ZFC set theory.

at least, that the background set theory is naïve, where there are fewer restrictions on set-theoretic constructions.[5] In summary, we have a situation where mathematicians knew about the paradoxes and they continued to use a known-to-be-inconsistent mathematical theory in the development of other branches of mathematics and in applications beyond mathematics.

This raises a number of interesting philosophical questions about inconsistent mathematics, its logic, and its applications. We will pursue some of these issues in this chapter. The first concerns the logic used in mathematics. It is part of the accepted wisdom that in mathematics, classical logic rules. Despite a serious challenge from the intuitionists in the early part of the twentieth century, classical logic is generally thought to have prevailed. But now we have a new challenge from logics more tolerant to inconsistency, so-called *paraconsistent logics*. In the next section I will give a brief outline of paraconsistent logics and discuss their relevance for the question of the appropriate logic for mathematics. Could it be that such 'deviant' logics are appropriate for mathematics?

The second general topic we will explore concerns applications of inconsistent mathematics, both within mathematics itself and in empirical science. There are many questions here, but I will focus on two: how can an inconsistent theory apply to a presumably consistent world; and what do the applications of inconsistent mathematical theories tell us about what exists? But before we broach such philosophical matters, I will first present a couple of examples of inconsistent mathematical theories.

7.1.2 An inconsistent mathematical theory

We have already seen Russell's paradox, the paradox arising from the set of all sets that are not members of themselves: $\mathcal{R} = \{x : x \notin x\}$. The paradox arises because of an axiom of naïve set theory known as *unrestricted comprehension*. This axiom says that for every predicate, there is a corresponding set. So, for example, there is the predicate 'is a cat' and there is the set of all cats; there is the predicate 'natural number' and there is the set of all natural numbers. So far, so good. The trouble starts when we consider predicates such as 'is a set' or 'is a non-self-membered set'. If there are sets corresponding to these two predicates, we get Cantor's cardinality paradox

---

[5] See Enderton (1977) for more on set theory.

and Russell's paradox, respectively. Cantor's cardinality paradox is derived by assuming that there is a set of all sets, Ω, with cardinality ω. Now consider the power set of Ω: $\mathcal{P}(\Omega)$. We have seen how Cantor's Theorem can be invoked to show that the cardinality of $\mathcal{P}(\Omega)$ is strictly greater than ω. But Ω is the set of all sets and so must have cardinality at least as large as any set of sets. Since $\mathcal{P}(\Omega)$ is a set of sets, we have a contradiction.

The naïve axiom of unrestricted comprehension was seen to be the culprit in all the paradoxes, and mathematicians set about finding ways to limit the scope of this powerful principle. One obvious suggestion is to simply ban the problematic sets – the set of all sets, the Russell set, and the like. This, however, is clearly ad hoc. Slightly better is to ban all sets that refer to themselves (either explicitly or implicitly) in their own specification. The generally agreed upon solution achieves the latter by invoking axioms that ensure that such problematic sets (and others) cannot be formed. This is ZFC. The basic idea is to have a hierarchy of sets, where sets can be formed only from sets of a lower level – a set cannot have itself as a member, for instance, because that would involve collecting sets from the same level. Nor can there be a set of all sets – only a set of all sets from lower down in the hierarchy. ZFC has not engendered any paradoxes, but it has the look and feel of a theory designed to avoid disaster rather than a natural successor to naïve set theory. More on this later.

Another important example of an inconsistent mathematical theory is the early calculus. When the calculus was first developed in the late seventeenth century by Newton and Leibniz, it was fairly straightforwardly inconsistent. It invoked strange mathematical items called infinitesimals (or fluxions). These items were supposed to be changing mathematical entities that approach zero. The problem is that in some places these entities behaved like real numbers close to zero but in other places they behaved like zero. Take an example from the early calculus: differentiating a polynomial such as $f(x) = ax^2 + bx + c$.[6]

$$f'(x) = \frac{f(x+\delta) - f(x)}{\delta} \tag{7.1}$$

---

[6] The omission of the limit '$\lim_{\delta \to 0}$' from the right-hand side on the first four lines of the following calculation is deliberate. Such limits are a modern development. At the time of Newton and Leibniz, there was no rigorous theory of limits; differentiating from first principles was along the lines presented here.

$$= \frac{a(x+\delta)^2 + b(x+\delta) + c - (ax^2 + bx + c)}{\delta} \tag{7.2}$$

$$= \frac{2ax\delta + \delta^2 + b\delta}{\delta} \tag{7.3}$$

$$= 2ax + b + \delta \tag{7.4}$$

$$= 2ax + b \tag{7.5}$$

Here we see that at lines one to three the infinitesimal δ is treated as non-zero, for otherwise we could not divide by it. But just one line later we find that $2ax + b + \delta = 2ax + b$, which implies that $\delta = 0$. The dual nature of such infinitesimals can lead to trouble, at least if care is not exercised. After all, if infinitesimals behave like zero in situations like lines four and five above, why not allow:

$$2 \times \delta = 3 \times \delta$$

then divide by δ to yield

$$2 = 3?$$

This illustrates how easily trouble could arise and spread if seventeenth- and eighteenth-century mathematicians weren't careful. There were rules about how these inconsistent mathematical objects, infinitesimals, were to be used. And according to the rules in question, the first calculation above is legitimate but the second is not. No surprises there. But one can quite reasonably ask after the motivation for the rules in question. Such rules about what is legitimate and what is not require motivation beyond what does and what does not lead to trouble.

The calculus was eventually, and gradually, made rigorous by the work of Bernard Bolzano (1781–1848), Augustin-Louis Cauchy (1789–1857), Karl Weierstrass (1815–97), and others (see Kline 1972) in the nineteenth century. This was achieved by a rigorous (ϵ-δ) definition of limit. More recently there has been a revival of something like the original infinitesimal idea by Abraham Robinson (1918–74) and John Conway (1937–), and even an explicitly inconsistent theory of infinitesimals by Australian philosopher Chris Mortensen.[7] So, to be clear, I am not claiming that there are any

---

[7] See Robinson (1966), Conway (1976), and Mortensen (1995).

ongoing consistency problems for the calculus. The point is simply that for over a hundred years mathematicians and physicists worked with what would seem to be an inconsistent theory of calculus.[8]

7.2 Paraconsistent logic

Classical logic has it that an argument form known as *ex contradictione quodlibet* or *explosion* is valid. The argument form was used in my proof of Fermat's Last Theorem at the beginning of this chapter. According to explosion any arbitrary proposition follows from a contradiction. The negation of Fermat's Last Theorem, or anything else, can be proven just as easily, and with pretty much the same proof as the one I opened with. Logics in which this argument form is valid are said to be *explosive*.[9] A *paraconsistent logic* is one that is not explosive. That is, in a paraconsistent logic at least one proposition does not follow from a contradiction. *Ex contradictione quodlibet* is invalid according to such logics.

7.2.1 The logic of paradox

There are many paraconsistent logics in the market place but let me sketch the details of one very straightforward such logic, just to make the discussion concrete. The logic LP (for 'the Logic of Paradox') is a three-valued logic with values $0, i$, and 1 (here 1 is 'true', 0 is 'false', and i is a third truth value). So far nothing unusual – quite a few logics have three values – but the interesting feature of this logic is that the crucial notion of validity is defined in terms of preservation of two of the truth values: an argument is valid if whenever the truth value of the premises are not 0, the truth value of the conclusion is not 0. Alternatively, we can define validity in terms of *designated truth values*: whenever the premises are designated, so too is the conclusion. Here, a designated value is one we want to be preserved in valid arguments. In a two-valued logic it is natural to designate 1 (or 'true'). But in a three-valued logic we can choose to designate 1 or we can designate both

---

[8] There are also cases where explicitly inconsistent, but non-trivial, theories have been developed. See Meyer (1976), Meyer and Mortensen (1984), Mortensen (1995), Priest (1997; 2000).

[9] Intuitionistic logic is also explosive.

1 and i. Designating both 1 and i, as we do in LP, is thus a natural extension of the usual definition of validity in classical logic: an argument is valid if whenever the premises are true, the conclusion is also true. Designating everything other than 0 does not matter in classical logic, since there are only two truth values (non-falsity and truth are the same thing). But in a three-valued logic, this makes all the difference. We also need to define the operator tables for the logical connectives (i.e., define how conjunctions, disjunctions, and negations get their truth values).[10] The operator tables for the logical connectives negation, conjunction, and disjunction (respectively) are as follows:[11]

$\neg$			$\wedge$	1	i	0		$\vee$	1	i	0
1	0		1	1	i	0		1	1	i	1
i	i		i	i	i	0		i	i	i	i
0	1		0	0	0	0		0	1	i	0

From these we see that if some sentence P has the truth value i, its negation, $\neg P$, also has the value i, and so does the conjunction of the two: $P \wedge \neg P$. Now take some false sentence Q (i.e., whose truth value is 0) and consider the argument from $P \wedge \neg P$ to Q. In LP this argument is invalid, since the premise $P \wedge \neg P$ does not have the truth value 0 and yet the conclusion Q does have the truth value 0. In this logic the 'proof' of Fermat's Last Theorem that I gave earlier is invalid. But also notice that if you restrict the truth values to 0 and 1, you get the same results as classical logic (just strike out the i columns and rows from the tables above and you recover the classical tables). So this logic is a conservative extension of classical logic, and a very natural extension at that.

---

[10] See Beall and van Fraassen (2003), Priest (2008), or Priest and Tanaka (2004) for full details and further discussion. The operator tables are the same as for the Kleene strong logic K_3, which will be familiar to logic students.

[11] These operator tables define negation ($\neg$), conjunction ($\wedge$), and disjunction ($\vee$) respectively. They are read as follows: (i) in the first table, read the right-hand column as giving the truth values of the unnegated proposition and the left-hand column as giving the corresponding truth value for the negation; (ii) in the second and third tables, read the top row and the left column (the ones separated from the main table by horizontal and vertical lines, respectively) to represent the truth values of the two conjuncts/disjuncts, and the corresponding entry of the main table gives the truth value of the conjunction/disjunction.

Think for a moment about the third truth value, i, in LP. How should we interpret this truth value? For a start, i is designated, so it behaves like truth. But the operator table for negation and conjunction reveals that if A has truth value i, $\neg A$ also has truth value i. Since i is designated, this gives us that $A \wedge \neg A$ is designated, whenever A takes truth value i. In light of this and related issues, a very natural interpretation of i in LP is as 'both true and false'.[12]

7.2.2 Reasons to be paraconsistent

What's the philosophical significance of all this? Well, it might just be that mathematicians were never tempted by the proof of Fermat's Last Theorem I gave earlier because the appropriate logic of mathematical proofs is a paraconsistent one. This sounds implausible, though. Surely all we need to do is ask a mathematician which logic they use and surely they'll all answer 'classical logic' (or perhaps 'intuitionistic logic'). For various reasons it might be interesting to conduct such sociological research of mathematicians' beliefs, but doing so will not help us answer the question at hand about the logic of mathematics. Our question is which logic do mathematicians *actually use*, and this is determined by mathematical practice, not by what mathematicians claim they use. Indeed, few mathematicians are experts in the differences between the various logics under consideration.

Here's another possibility. Perhaps mathematicians don't use a paraconsistent logic but, rather, just avoid proofs like the five-line proof of Fermat's Last Theorem given earlier. Indeed, they might steer clear of contradictions generally. The latter is hard to do, though, when you're working in a known-to-be-inconsistent theory. But perhaps part of what it takes to be a good mathematician is to recognise not just valid proofs, but also sensible ones. On this suggestion the proof I opened with might be formally valid but it's not sensible, since it involves a contradiction (it takes a contradiction as a premise). But this won't do as a response. First, the contradiction in question can be proven straightforwardly in a very rigorous way from what was, at the time, the best available theory of sets; it's not some

---

[12] Logic students should compare this third truth value with the third truth value in the logic K_3. In the latter the third truth value is usually thought of as 'neither true nor false' because it is not designated.

implausible proposition without any support. Second, not all arguments involving contradictions (or taking contradictions as premises) are defective. Take the argument from $P \wedge \neg P$ therefore $\neg P$. Surely this is both valid and sensible.[13] Putting these issues aside, the most serious problem with this line of response is that the notion of a sensible proof is in need of clarification. The advocate of a paraconsistent logic has no such problems here; they have only the one notion: (paraconsistent) validity, and the proof in question fails to be valid.

Even if mathematicians do use classical logic but exercise some (ill-defined) caution about what proofs to accept above and beyond the valid ones, perhaps the practice in question can be modelled using a paraconsistent logic. Or even stronger still, perhaps the best practice is appropriately modelled with a paraconsistent logic. If so, we might suggest that mathematicians *ought* to use a paraconsistent logic. As I've already suggested, one reason for embracing paraconsistent logic is that it provides a very natural way to block the undesirable proofs. But there are other reasons to entertain a paraconsistent logic. There are many situations in mathematics where the consistency of a theory is called into question but without a demonstration of any inconsistency. Consider, for example, the earliest uses of complex numbers, numbers of the form $x + yi$, where $i = \sqrt{-1}$ and x and y are real numbers. There was a great deal of debate about whether it was inconsistent or just weird to entertain the square root of negative numbers.[14] Moreover, it was not just the status of complex analysis that was at issue. If the theory of complex analysis turned out to be inconsistent, everything that depended on it, such as some important results in real analysis, would also be in jeopardy. Adopting a paraconsistent logic is a kind of insurance policy: it stops the rot from spreading too swiftly and too far – whether or not you know about the rot.

Perhaps the most interesting reason to entertain a paraconsistent logic in mathematics is that with such a logic at hand, naïve set theory and naïve infinitesimal calculus can be rescued (Mortensen 1995). There is no need to adopt their more mathematically sophisticated replacements: ZFC and modern calculus. There are a couple of pay-offs here. First, both naïve set

---

[13] Why? This argument is simply an instance of simplification or conjunction elimination: $A \wedge B$ therefore B.

[14] See Kline (1972) for some of the relevant history of this debate.

theory and naïve infinitesimal calculus are easier to teach and learn than their modern successors. In naïve set theory there is no need to deal with complicated axioms designed to block the paradoxes; the easily understood and intuitive unrestricted comprehension is allowed to stand. With naïve calculus there is no need to concern oneself with the subtle modern ϵ-δ definition of limit; infinitesimals are allowed back in the picture.[15] The second pay-off is related to the first and concerns the intuitiveness of the theories in question. At least in the case of set theory, the naïve theory is more intuitive. ZFC, for all its great power and widespread acceptance, remains unintuitive and even ad hoc. There is no doubt that naïve set theory is the more natural theory. Similar claims could be advanced in relation to naïve infinitesimal calculus over modern calculus, though admittedly the case is not as clear here.

7.3 Applying inconsistent mathematics

Let's now turn to the application of inconsistent mathematics. There are many interesting issues here, and I'll say just a little about a few of these. The first issue is that inconsistent mathematics adds a new twist to the 'unreasonable effectiveness of mathematics', which we looked at in Chapter 6. Recall that the puzzle is to explain how an a-priori discipline such as mathematics can find applications in a-posteriori science.

One proposal for explaining away the apparent puzzle was the mapping account of mathematical applications. According to this account there is a mapping between the mathematical structure and the physical system being modelled. This mapping tracks the structural similarities between the two and this is enough to explain why mathematics is so useful. Mathematicians develop structures often motivated by, or at least inspired by, physical structures. The mathematician's structures then (unsurprisingly) turn out to be similar to various physical structures. But the fact that inconsistent mathematics, such as the early calculus, finds wide and varied applications in empirical science raises problems for this line of thought. After all, assuming, as most of us do, that the world is consistent, how can an inconsistent mathematical theory be similar in structure to something that's

---

[15] As they are in non-standard analysis, but non-standard analysis does this in a different way (Conway 1976; Robinson 1966).

consistent? There is a serious mismatch here. It certainly cannot be that the inconsistent mathematics in question is isomorphic to the world, unless one is prepared to countenance the possibility that the world itself is inconsistent. I'm not suggesting that the above thought about how to dissolve the puzzle of the unreasonable effectiveness of mathematics is completely off the mark, just that it cannot be the whole story.

The second issue in relation to applying inconsistent mathematics takes us back to metaphysics. Recall from Chapter 3 that the indispensability argument pushes for belief in the reality of mathematical objects, from the fact that mathematical theories are indispensable to our best scientific theories. Again, applications of inconsistent mathematics add a new twist. There have been times when inconsistent mathematical theories (most notably the early calculus) have been indispensable to a broad range of scientific theories. Seventeenth- and eighteenth-century calculus was indispensable to mechanics, electromagnetic theory, gravitational theory, heat conduction, and much more. It seems that if one subscribes to the indispensability argument, then there's a rather unpalatable conclusion beckoning: sometimes we ought to believe in the existence of inconsistent objects. For example, the early calculus posited inconsistent objects: infinitesimals, which were taken to be both equal to zero and not equal to zero.[16]

It is not clear what to make of this argument for the existence of inconsistent objects. Is it a *reductio* of the original indispensability argument? Does it tell us that consistency should be an overriding constraint in such matters? If so, on what grounds? Perhaps it is not as crazy as it sounds to believe in inconsistent mathematical objects. It is fair to say that the jury is still out on these issues, with much more work and detailed examination of case studies required before a verdict will be delivered.

Finally, there has been some fascinating work on using inconsistent mathematical theories – more specifically, inconsistent geometry – to model inconsistent pictures such as those of M. C. Escher (1898–1972) and Oscar Reutersvärd (1915–2002) (e.g., Escher's Belvedere and Reutersvärd's Penrose triangle). Mortensen (1997) has argued convincingly that consistent mathematical theories of such pictures do not do justice to the cognitive dissonance associated with seeing such pictures *as impossible*. Arguably, the

---

[16] In general, we can think of an inconsistent object, in the sense intended here, as an object which is taken to have contradictory properties.

dissonance arises from the perceiver of such a picture constructing an inconsistent mental model of the situation – an impossible spatial geometry. Any consistent mathematical representation of this inconsistent cognitive model will fail to capture its most important quality, namely its impossibility. Inconsistent mathematics, on the other hand, can faithfully represent the inconsistent spatial geometry being contemplated by the perceiver and thus serve as a useful tool in exploring such phenomena further. These applications of inconsistent mathematics should hold interest beyond philosophy. Indeed there are immediate applications in cognitive science and psychology. But such work is very new, and its full significance has not yet been properly appreciated.

Inconsistent mathematics has received very little attention in mainstream philosophy of mathematics and yet, as we have seen here, several interesting philosophical issues are raised by it. Moreover, some of these issues – such as the ontological commitments of inconsistent mathematical theories and the use of paraconsistent logic as the logic for mathematics – bear directly on contemporary debates in philosophy of mathematics. Other issues – such as the application of inconsistent mathematics to model inconsistent pictures – promise to take philosophy of mathematics in new and fruitful directions. For my money, though, the biggest issue concerns possible insights into the relationship between mathematics and the world. This is a central problem for both philosophy of mathematics and philosophy of science. I believe that there is a great deal to be learned about the role of mathematical models – both consistent and inconsistent – in scientific theories by paying closer attention to the use of inconsistent mathematics in applications. Perhaps focusing our attention on the consistent mathematical theories has misled us to some extent. If this is right, we won't have the complete picture of the mathematics–world relationship until we understand how inconsistent mathematics can be so useful in scientific applications.

Discussion questions

1. Was the early calculus inconsistent? Perhaps it was just underdeveloped and perhaps ambiguous. If it was inconsistent, but not recognised to be inconsistent, does that make its indispensability any more acceptable?

2. Can it ever be rational to believe a contradiction? What if you have good reason to believe P and good reason to believe Q, where Q entails $\neg P$?

3. Can you think of other reasons for embracing a paraconsistent logic in one's reasoning?

4. Show that the following logical truths of classical logic are also logical truths in LP (i.e., always take a designated value):
   - $P \vee \neg P$ (excluded middle)
   - $\neg(P \wedge \neg P)$ (law of non-contradiction).

5. Let '$\supset$' be the material conditional, where $P \supset Q$ is defined as $\neg P \vee Q$. Construct an operator table for this connective. Show that modus ponens is invalid for this connective in LP. Show that pseudo modus ponens, $(P \wedge (P \supset Q)) \supset Q$, is a logical truth, though.

6. Is it possible to distinguish a consistent object posited by an inconsistent theory from an inconsistent object posited by an inconsistent theory? If so, might we be able to argue that although the early calculus was inconsistent, the mathematical entities – infinitesimals and the like – were nevertheless consistent?

7. Suppose you're using an inconsistent mathematical theory to model some physical system. Can you simply deny that the inconsistent parts of the mathematics represent anything real? Perhaps the inconsistent parts of the theory can be viewed as mere artefacts of the model. This would make inconsistent mathematical models much like other scientific models with false assumptions. But how do you isolate the false assumptions in the inconsistent mathematics? Is this going to be more difficult if the logic employed is explosive?

8. Explain why one might prefer an inconsistent theory of the geometry of Escher drawings.

Recommended further reading

For a good account of the inconsistencies in set theory at the beginning of the twentieth century, see Giaquinto (2002). For further details of paraconsistent logics, see Beall and van Fraassen (2003), Priest (2008), and Priest and Tanaka (2004). For more on the philosophical upshot of applications of inconsistent mathematics, see Colyvan (2008; 2009) and Mortensen (2004). For the technical details of explicitly inconsistent mathematical theories, see Meyer and Mortensen (1984) and Priest (1997). For the debate over a classical or an inconsistent geometry for impossible objects, see Penrose and Penrose (1958) and Mortensen (1997).

Beall, J. C. and van Fraassen, B. C. 2003. *Possibilities and Paradox*, Oxford University Press.

Colyvan, M. 2008. 'The Ontological Commitments of Inconsistent Theories', *Philosophical Studies*, 141(1): 115–23.

  2009. 'Applying Inconsistent Mathematics', in O. Bueno and Ø. Linnebo (eds.), *New Waves in Philosophy of Mathematics*, Basingstoke, UK: Palgrave Macmillan, pp. 160–72, reprinted in M. Pitici (ed.), *The Best Writing on Mathematics 2010*, Princeton University Press, 2011, pp. 346–57.

Giaquinto, M. 2002. *The Search for Certainty: A Philosophical Account of Foundations of Mathematics*, Oxford University Press.

Meyer, R. K. and Mortensen, C. 1984. 'Inconsistent Models for Relevant Arithmetic', *Journal of Symbolic Logic*, 49: 917–29.

Mortensen, C. 1997. 'Peeking at the Impossible', *Notre Dame Journal of Formal Logic*, 38(4): 527–34.

  2004. 'Inconsistent Mathematics', in E. N. Zalta (ed.), *The Stanford Encyclopedia of Philosophy* (Fall 2004 edn), http://plato.stanford.edu/archives/fall 2004/entries/mathematics-inconsistent/.

Penrose, L. S. and Penrose, R. 1958. 'Impossible Objects, a Special Kind of Illusion', *British Journal of Psychology*, 49: 31–3.

Priest, G. 1997. 'Inconsistent Models of Arithmetic Part I: Finite Models', *Journal of Philosophical Logic*, 26(2): 223–35.

  2008. *An Introduction to Non-Classical Logic: From If to Is*, 2nd edn, Cambridge University Press.

Priest, G. and Tanaka, K. 2004. 'Paraconsistent Logic', in E. N. Zalta (ed.), *The Stanford Encyclopedia of Philosophy* (Winter 2004 edn), http://plato.stanford.edu/archives/win2004/entries/logic-paraconsistent/.

8   A rose by any other name

*It is impossible to be a mathematician without also being a poet in spirit ... It
seems to me that the poet must see what others do not see, see more deeply than
other people. And the mathematician must do the same.*

Sophie Kowalevski (1850–91)[1]

One often hears the claim that mathematics is 'the language of science'.
This is meant as a compliment to mathematics. But mathematics is not the
language of science in the way that French is the language of love. The latter
is surely conventional, perhaps driven by aesthetic preferences for 'amour'
over 'love' and 'belle' over 'beautiful' and the like. In any case, mathematics,
if it is the language of science, is not like this. It's not as though science
looks or sounds sexier when it's written mathematically (actually, perhaps
it does, but that's by the by). The point of the slogan is to emphasise that
a great deal of science – especially physics, but many other branches of
science as well – is typically highly mathematical. Moreover, a great deal of
science could not even be formulated without mathematics.

In this chapter I will argue that although there is undoubtedly something
right about the view of mathematics as the language of science, it seriously
undersells mathematics. To think of mathematics as *merely* the language of
science fails to appreciate the variety of roles mathematics plays in many
diverse branches of science. Thinking of mathematics as a language is useful in appreciating the significance of, and the difficulties encountered in
developing, a good notational system. Good notation is far from trivial. So
let's start by looking at some of the benefits of good notation. Along the

[1] S. Kovalevskaya, *A Russian Childhood*, trans. and with an Introduction by B. Stillman,
New York: Springer, 2010, p. 35. Reproduced with kind permission from Springer
Science+Business Media B.V.

way, we will see the role good notation can play in prompting new ideas and new developments in mathematics and science.

8.1 More than the language of science

8.1.1 The natural numbers

Mathematicians and historians of mathematics have long recognised the importance of good notation in mathematics. But since the demise of formalism as a philosophy of mathematics, mathematical notation has received little philosophical attention.[2]

Let's start with a couple of examples. First, consider the Arabic notation for the natural numbers: $1, 2, 3, 4, 5 \ldots$ The familiarity of this notation system makes it easy to overlook just how powerful and extraordinary it is. James Robert Brown (2008, p. 85) notes that one of the reasons this notation is so powerful is that the most important feature of the natural numbers – their recursiveness – is built into the notation.[3] He suggests that '[t]his is the mathematical equivalent of poetry's onomatopoeia' (Brown 2008, p. 86). He goes on to suggest that the recursive nature would need to be known in advance of the invention of such notation. But this doesn't seem right. It is surely at least possible to conceive of a society without the concepts of infinity or of recursion. This society might still entertain the idea of a large but finite numbers system – enough numbers to make the Arabic notation worthwhile. One could even imagine a member of such a society noting that their notation allows for the representation of numbers larger than any they currently use. This, in turn, might lead to the discovery of the natural numbers in all their recursive glory. In short, Brown is too quick to dismiss the idea that the notation can help reveal hitherto unknown mathematical facts. At least, there is nothing in Brown's discussion to rule this out, and I suspect he would be friendly to the idea of such notation-driven discovery.

---

[2] One notable exception is Brown (2008), who has a whole chapter on the philosophical significance of mathematical notation.

[3] The notation also makes it easy to assess many inequalities. For example, it is obvious that 373,559 is greater than 4,749; all one needs to do is note that the first string is longer than the second. Notice that this is not the case in Roman notation, where, for instance x > viii.

Another important feature of good notation is that it can facilitate computation. The Arabic notation delivers all the standard algorithms for addition, subtraction, multiplication, and division that we learned in primary school. It does this, of course, by appreciating how the recursive structure of the natural numbers is represented in the Arabic notation. This is far from trivial. Try multiplying two large numbers directly in Roman numerals and you will soon see the superiority of Arabic notation.[4] It is not always made clear that such algorithms depend on the notation. This is something that the formalists were well aware of: you can reduce a great deal of mathematics to manipulations of particular symbols. Recall (from Chapter 1) that the formalists saw such computation via symbol manipulation as the primary activity of mathematics. Be that as it may, there does seem to be calculational power in at least some notational systems.[5]

8.1.2 Elementary differential calculus

Now consider the notation of elementary calculus. It is well known that Leibniz and Newton had different notation for the derivatives of functions. Leibniz used dy/dx for the first derivative and d^2y/dx^2 for the second derivative, whereas Newton used the dot (or prime) notation: $\dot{f}$ and $\ddot{f}$ (or f' and f'') respectively. Newton's notation is more economical, but it does not generalise so well to higher dimensions, where one needs to be explicit about which independent variable we are differentiating with respect to. Indeed, Newton was mostly interested in differentiating with respect to time (and the dot notation is still commonly used for time derivatives). Leibniz's notation generalises very well to higher-order partial derivatives, because

---

[4] Brown (2008, pp. 86–93) gives other instructive examples of computational power arising from notation. For example, he considers knot theory, where the notation allows for the determination of whether two tangles are the same. See especially the result invoking Conway notation and continued fractions.

[5] The Arabic notation also enables some cool party tricks – well, they're cool at a certain kind of nerdy party! Think of a number between 1 and 9, multiply it by 3, subtract 2, add your age when you first tasted Chinese food, add 0 if you've never tasted Chinese food, add the number of siblings you have, multiply by 9, than add the digits of the resulting number, repeatedly, if necessary until you arrive at a number between 1 and 9. The result is 9, right? The crux of this trick is that any multiple of 9 has its digits sum to 9 or else sum to a number whose digits sum to 9, or whose digits sum to a number whose digits sum to 9, and so on. This result falls out of modular arithmetic, but it involves the notation in a non-trivial way.

even in the case of one independent variable, the notation is explicit about differentiating with respect to the variable in question.

I am not claiming any victory for Leibniz over Newton here; it's just that this again looks like a case where good notation can facilitate new mathematics (in this case multi-variate calculus) by making the transition to the general case seamless. This example is different from the last in that here we see that notation might be good for one purpose but not for another. Also note how once again good notation might suggest new mathematics. In Leibniz notation, it is very natural to consider the question of whether the mixed partial derivatives $\partial^2 u/\partial x\partial y$ are the same as the other mixed partial derivatives $\partial^2 u/\partial y\partial x$.[6] It turns out that these mixed partial derivatives are equal at any point where the function u has continuous mixed partial derivatives. For present purposes we simply note that questions such as this (and the theorem that answers this question) naturally arise from the notation itself. Good notation, it seems, prompts the user to keep track of distinctions the inventor of the notation may not have even noticed.

Next, consider the standard notation for the Laplace operator in two dimensions:

$$\Delta f(x, y) = \frac{\partial^2 u}{\partial x^2} + \frac{\partial^2 u}{\partial y^2}. \tag{8.1}$$

This notation is economical, allowing us to write Laplace's equation in two dimensions as: $\Delta u(x, y) = 0$, and if the dimension is understood, we need only write: $\Delta u = 0$. The economy of this notation is even more apparent when we go beyond two dimensions. The generalised n-dimensional Laplace operator is defined as

$$\Delta u = \sum_{i=1}^{n} \frac{\partial^2 u}{\partial x_i^2} = \frac{\partial^2 u}{\partial x_1^2} + \frac{\partial^2 u}{\partial x_2^2} + \frac{\partial^2 u}{\partial x_3^2} + ... \frac{\partial^2 u}{\partial x_n^2}. \tag{8.2}$$

It is worth pointing out that the notation itself, in so far as it ignores the dimension, might even be thought to suggest the generalised n-dimensional version of the operator. Even if the generalisation were not already in play, the notation would remind one that there is nothing special about two (or any other number of) dimensions. We thus see that good notation can be economical as well as lend itself to further mathematical developments.

---

[6] The former is the derivative taken with respect to x and then with respect to y, whereas the latter is in the other order.

Since mathematical developments are very often generalisations to more abstract structures, we can restate this last virtue of good notation as that of *facilitating more abstract, generalised mathematical theories*.[7]

8.1.3 Topology

Next consider an example from topology. Topology is the study of features of spatial structures preserved under continuous transformations, or as the slogan goes: topology is 'rubber-sheet geometry'. To get a feel for this fascinating area of mathematics, we can transform a solid sphere into a solid cube without tearing or introducing holes, or filling in holes. (Think of a solid sphere made of clay and how it could be moulded into a solid cube.) From the point of view of topology, a cube and a sphere are identical. But a torus (the shape of an American doughnut) is not the same as either of these – the hole makes all the difference here. It is the hole that prevents any continuous transformation of a sphere, say, into a torus. At some stage a hole must be punched through the sphere (i.e., a discontinuity in the transformation is necessary). A torus, however, is topologically equivalent to a teacup (because of the teacup handle).

There is a particularly powerful piece of diagramatic notation that's a kind of cardboard cut-out plan – except that the cardboard is flexible and the construction sometimes cannot be completed in the space we occupy. Let's start with a couple of familiar ones to get the idea. A cylinder is represented thus:

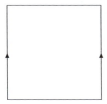

---

[7] There is also the related issue of which concepts are the most fruitful ones to focus on. Imre Lakatos's most significant contribution to the philosophy of mathematics was to argue (convincingly) that mathematical concepts are very often refined in the light of mathematical results and counterexamples (Lakatos 1976). So we might reasonably ask why the concept of the Laplacean operator is so useful in both pure and applied mathematics. For the moment I set this interesting issue aside and assume that we have already decided which mathematical concepts are useful and are concerned with the question of representing them in our mathematical notation. We revisit the issue in section 8.3.

The arrow facing upwards on the left side and the arrow pointing upwards on the right indicate that these two edges are identified. That is, a point on the right edge, for the purposes of present interests, is the same point as the corresponding point on the left edge. The fact that the two arrows are facing in the same direction means that identification is made with the same orientation. That is, the top left corner is identified with the top right corner, and the bottom left corner is identified with the bottom right corner, and so on for all the points along the edges. To give a more concrete interpretation of the notation, we can see it as instructions for building a cylinder: cut out the square of rubber or cardboard, or whatever, and glue the left edge to the right edge with the same orientation.

The orientation needn't be the same, of course. We might, instead, identify the two edges with reverse orientation thus:

Here the left and right edges are identified so that the top right is identified with the bottom left and the bottom right with the top left. That is, before gluing our rubber sheet, we put a twist in it so that we form a Möbius strip (named after its discoverer, the German mathematician August Möbius (1790–1868)). These are very interesting objects. They are two-dimensional surfaces but we need three dimensions to construct one of them. The twist needs to run through the third dimension.

Now using this notation we represent a torus.

Here we identify the right edge with the left edge, with the same orientation, as we did with the cylinder, but now we also identify the top edge with the bottom edge, again with the same orientation. That is, cut out a square sheet of rubber and glue the right edge to the left to form a cylinder, then glue the two circular ends of the cylinder (what were the top and bottom edges of the original square) together to form a torus. (It doesn't matter whether we glue the sides first and then the edges or vice versa.)

Finally we use this notation to represent a very strange object indeed: the Klein bottle (named after its discoverer, German mathematician Felix Klein (1849–1925)).

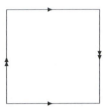

This is the same as the torus except that the second identification is performed with the reverse orientation. But think for a moment how you would use this diagram as instructions for building a Klein bottle. Glue the top and bottom edges as with the torus. So far so good. But now you're required to put in a twist to get the reverse orientation for the second gluing. You cannot physically do this because you've run out of dimensions. The twist needs to be in a fourth dimension. Since we're stuck with only three spatial dimensions, the Klein bottle is not constructible. It is a three-dimensional object needing four spatial dimensions for its construction. Perhaps its most interesting feature, however, is that it has no inside or outside.

The Klein bottle is a lovely demonstration of the power of good notation. The notation just developed makes it plain that there are such things as three-dimensional objects that require more than three spatial dimensions for their construction. The notation also shows that there are three-dimensional objects with no inside or outside. Before seeing and understanding the relevant notation, you'd be sorely tempted to dismiss the idea of such objects as nonsense. But even if you had not stumbled onto such things already, the notation just developed leads you to them, and we can even deduce the properties of objects such as the Klein bottle by studying the details of the notation. (For example, from the notation alone it can be seen that the Klein bottle has no inside or outside.) Arguably, the notation also helps us see why we cannot build such objects: because we are stuck in three spatial dimensions yet we need to engineer a twist in a fourth dimension.

Another interesting issue raised by this particular piece of notation involves the difference between algebraic methods and geometric methods. For a long time it has been the orthodoxy in mathematics to see geometric intuitions as unreliable. Instead, we should trust only purely algebraic methods. The latter are the epitome of rigour. Pictures and geometric intuitions are at best pedagogical devices, or so this line of thought goes. This view has been challenged recently (Brown 2008) and some have even suggested that the distinction between algebraic methods and geometric ones is not as sharp as it might first seem. This example lends support to the latter thesis. This algebraic topology notation is something of a halfway house between pure algebra and pure geometry. It is both notation and a kind of blueprint for construction of the objects in question. The first seems to belong to algebra, while the second is geometric. But whichever way you look at it, we have a powerful piece of notation here that does some genuine mathematical work for us.

8.1.4 The point at infinity

One final example of how clever notation can reveal something mathematically interesting. Here we see that one can add infinity to the complex plane, to deliver the extended plane, without leading to trouble. This is done by a cunning construction called *the stereographic projection*. But this construction can also be considered an alternative notation for the complex plane. Here's how it goes. Consider the complex plane laid out horizontally and with each point represented with the usual Cartesian 2-coordinates (x, y). Now add a third, vertical dimension, the z dimension, and with the usual Cartesian 3-coordinates (x, y, z). Consider a unit sphere with its centre at the origin $(0, 0, 0)$: $x^2 + y^2 + z^2 = 1$. Now we consider the line generated by joining a point in the x–y plane with the north pole of the hemisphere $(0, 0, 1)$. In particular, we are interested in where these lines intersect our unit sphere. For example, the line joining the point $(1, 0)$ in the x–y plane and the north pole of our sphere intersects the sphere at $(1, 0, 0)$. The line joining the point $(0, 0)$ to the north pole intersects the sphere at the south pole $(0, 0, -1)$. If we now identify each point of the Cartesian plane with the corresponding point of intersection of the sphere, we see that all the points inside the unit circle $x^2 + y^2 = 1$ are mapped to unique points on the part of the sphere below the x–y plane, all the points on the unit circle get mapped to their three-dimensional counterpart (i.e., (x, y) gets mapped to $(x, y, 0)$), and all the points outside the unit circle get mapped to unique points on the portion of the sphere above the x–y plane. Moreover, the further from the origin the point is, the closer its corresponding point of intersection is to the north pole of the sphere. Via this construction – a conformal (or angle-preserving) mapping – we have created an alternate representation for each point in the real plane.

What is useful about this apparently cumbersome representation is that every point on the plane is represented by an ordered triple (points on the surface of the sphere) and every point of the surface of the sphere, except one – the north pole – has a corresponding point on the x–y plane. But now we just add the north pole and stipulate that it is the point at infinity. The motivation for this should be plain to see. It should also be plain to see that the north pole represents infinity in all directions on the x–y plane. Using this construction (or alternative notation) we have shown that it is sensible to talk of the extended real plane (i.e., the complex plane with a single point at infinity added). That this can be done is not at all apparent in the standard notation. Again, we have found an interesting extension of our mathematics, and this extension was facilitated by the alternative notation employed in the stereographic projection.

8.2 Shakespeare's mistake

So far we have seen that good notation can enjoy the virtues of economy, calculational power, and facilitating advances in mathematics. But let's push things a little further. I want to pursue the suggestion, hinted at in the topology example, that good notation can make mathematical explanations more perspicuous.

Recall that in Chapter 5 we looked at mathematical explanations, both within mathematics and in empirical science. Here we focus on the former, intra-mathematical explanations. Let's take seriously the idea that mathematical explanation can facilitate explanation, as the Klein bottle example at least suggested.

8.2.1 Analytic geometry

Chief among René Descartes's (1596–1650) many contributions to philosophy, mathematics, and science, is his combining geometry and algebra in *analytic geometry*. The idea is so familiar now that it is hard to fully appreciate just how brilliant this innovation was. The idea is simply that we invoke a coordinate system for the plane, and then geometric figures can be represented algebraically in terms of these coordinates. In the other direction, we can represent algebraic equations and inequalities geometrically. The convention is to call the horizontal coordinate axis the x-axis and the vertical axis, the y-axis. For example, the circle with its centre at the origin of the coordinate system (the point $(0,0)$) and with radius 1 unit is written algebraically as $x^2 + y^2 = 1$. A parabola with its vertex at the origin, with focus at $(0,1)$ is $y = x^2/4$. This simple idea is very ingenious, and extremely powerful. With this algebraic notation for geometric figures, we are able to use geometry to help visualise otherwise abstract algebraic problems, and to use the rigorous methods of algebra to solve geometric problems.

For example, from the relevant geometry we can see why the polynomial $x^2 - 1$ has two real roots: because the corresponding parabola $y = x^2 - 1$ has its vertex below the x-axis. We can also see why the polynomial $x^2 + 1$ has no real roots: because the corresponding parabola $y = x^2 + 1$ sits entirely above the x-axis.[8] It is, of course, possible to understand why $x^2 + 1$ has no real roots by considering the algebra alone, but with the connection to geometry in place, you can *see* why. This is just the tip of the iceberg. By invoking the power of analytic geometry we can visualise differentiation (as the function representing the slope of a tangent to a curve at a given point), integration (as the function representing the area under a curve), and much more besides. These and many other applications should be very familiar, so we won't dwell on them here. It is worth reflecting for a moment, however, on the important role the algebraic notation for geometric figures plays in all this.

The preceding examples highlight the power of geometry when brought to bear on algebraic problems. But, for present purposes, we're more interested in the other side of the coin: the power of algebraic notation when brought to bear on geometry. A good example of this is the discovery of the Weierstrass function, named after the German mathematician Karl Weierstrass. It is tempting to think that continuous functions can fail to

---

[8] Although the fundamental theorem of algebra guarantees that the polynomial in question has two complex roots: $\pm i$, where $i = \sqrt{-1}$.

be differentiable at most at a countable number of points.[9] That is, if a function is continuous on an interval, it may have a countable number of points at which it is not differentiable, but differentiable at the rest. For example, the function

$$f(x) = \begin{cases} x + 1, & x < 0 \\ -x + 1, & x \geq 0 \end{cases}$$

is continuous everywhere but is not differentiable at the point $x = 0$; there is a cusp at $x = 0$. Weierstrass managed to show that there are continuous functions that are nowhere differentiable – they have cusps everywhere. The function Weierstrass originally produced was

$$f(x) = \sum_{n=0}^{\infty} a^n \cos(b^n \pi x),$$

where $0 < a < 1$, b is a positive odd integer and $ab > 1 + 3\pi/2$. There are many other such functions. It is very hard to imagine discovering such functions without algebraic notation in geometry. After all, pure geometric intuitions, if anything, suggest that there are no such functions. But more importantly, the algebraic notation allows the explanation of why some continuous functions may fail to be differentiable.

My suggestion in this and the following example is that the notation helps us understand what's going on, and in this sense helps engender (psychological) explanations. By this I don't meant to suggest that the notation is the reason for the phenomenon in question, just that the notation helps make the explanation accessible to us.[10] Next we consider an example where algebraic notation is used in a more essential way, and the explanation in question seems to rely on the algebraic representation of some geometry.

---

[9] Recall from section 2.1 that a countable set is either finite or can be placed in one–one correspondence with the natural numbers.

[10] We can distinguish two senses of explanation: one an objective sense and the other a psychological sense. For example, a scientific explanation might not be understandable to someone not versed in the relevant science. That does not stop it from being an explanation in the objective sense, but its failure to enlighten the person in question prevents it from being a psychological explanation for that person. It is the latter, psychological sense of explanation we're concerned with here.

8.2.2 Squaring the circle

There is an ancient problem – going back to the Greeks – of constructing a square of the same area as a given circle, using just a straightedge and compass.[11] That is, using only these two instruments we must construct a square of side $r\sqrt{\pi}$, where r is the radius of the circle in question and π is the ratio of the circumference to the diameter of a circle. After centuries of attempts, the impossibility of this construction was finally proved in 1882 by Ferdinand von Lindemann (1852–1939). The proof, however, comes from abstract algebra, not geometry. A quick sketch of the connection here is worthwhile, because the algebraic notation is crucial to the proof in question.

First, we catalogue the legitimate, basic ruler and straightedge constructions (drawing a line through two existing points, constructing a circle with its centre at one existing point and running through another existing point, and so on). We then provide notation for the basic geometric objects (lines, points, and arcs of the circle) and note that we can represent these objects in the Cartesian plane, in the usual way. We then show that the permissible geometric constructions give rise to a small set of algebraic operations on line lengths: addition, subtraction, division, multiplication, and taking the square root. The idea here is that if two line segments of lengths a and b are given, we can construct line segments of length ab, $a + b$, $a - b$, a/b, and $\sqrt{a}$. What is crucial is that these are the *only* algebraic operations the geometric constructions license.[12]

Note that what we have done is again forge a link between the geometric constructions and algebra, and, in particular, we can now employ algebraic methods. We have thus transformed the problem from a geometric construction of a given length, $\sqrt{\pi}$, to a problem in algebra of determining whether $\sqrt{\pi}$ is obtainable from a given number by successive applications of the algebraic operations just listed. Thus stated, the problem reduces to whether $\sqrt{\pi}$ can be the root of a polynomial with rational coefficients and where the only powers are 0, 1, or an even integer. Now

---

[11] This is one of three famous ancient geometric construction problems, the other two being trisecting an angle and doubling the cube. The first is the problem of trisecting an arbitrary angle, the second is that of creating a cube with twice the volume of a given cube.

[12] The latter is not obvious and requires proof.

this is where the Lindemann result comes in. Lindemann proved that π (and therefore $\sqrt{\pi}$) is transcendental. That is, π (and $\sqrt{\pi}$) is not the root of any polynomial with rational coefficients. Squaring the circle is thus impossible.[13]

Although, historically and mathematically, the fact that π is transcendental is the key to the impossibility result, it is important to see how the problem needed to be set up as an algebraic problem. This involved the clever introduction of algebraic notation for the geometric objects and operations, and noting that the geometric operations give rise to some familiar algebraic systems. Again, we see good mathematical notation playing a key role in delivering a mathematical explanation. (For the explanation of the impossibility here is the transcendentalness of π.)

We have seen that good notation can enjoy the virtues of economy, calculational power, and facilitating advances in mathematics. More controversially, good notation may also contribute to mathematical explanations (as in some of the topology examples). This already takes us well beyond the standard view (in so far as there is such a thing) of mathematical notation. The standard view I have in mind here suggests that it's the mathematical objects that matter, not the notation we use for them: as Shakespeare put it in *Romeo and Juliet*

> What's in a name? that which we call a rose
> By any other name would smell as sweet.

Following Shakespeare's lead, it might be tempting to suggest that a mathematical object by any other name would be just as useful. But we have seen that this is not so. Sometimes the names encode properties of the objects in question, and in such cases other names would be less revealing.[14] Perhaps because of this encoding, good notation can keep track of distinctions we may not have initially noticed and force us to investigate such

---

[13] Other classic geometric construction problems – trisecting an angle and doubling the cube – also turn out to be impossible. They are impossible for different reasons: respectively, $\cos(\pi/9)$ is not the root of any of the polynomials just described, nor is $\sqrt[3]{2}$. In each case, the same notation outlined here is used and is important in delivering the impossibility result in question.

[14] Interestingly, this is true of natural language as well: onomatopoeias reveal something about the sounds they name, and other words such as 'computer' and 'amplifier' tell you something about the objects they name.

distinctions and see possibilities for future research not previously antici-
pated. In short, properties of the notation are important in mathematics. So
Shakespeare was wrong, at least about mathematical notation, but I'd suggest he was wrong about natural language as well. A rose by any other name
may well smell as sweet, but still some names are more revealing about that
which they name. In any case, it is far from clear that mathematics would
be served equally well by alternative notations. Getting the notation right
features prominently in mathematical practice. And there is a good reason
for this. Good notation does serious work in mathematics. There is a great
deal more philosophical work to be done on understanding and appreciating the role of notation in the various branches of mathematics. Especially
important here is the development of an account of how good notation can
advance mathematics and contribute to mathematical understanding.

8.3 Mathematical definitions

Let's return to the issue of the mathematical concepts behind the notation.
How do mathematicians decide which mathematical concepts to study and
to feature in their theorems? They are free to study whichever concepts they
like, but some will turn out to be more interesting and useful than others.
Sometimes the definitions need to be revised because they lead to trouble.
The implicit definition of 'set' delivered by naïve set theory needed to be
revised in the light of the set-theoretic paradoxes. In effect, the various set
theories we currently have at our disposal can be seen as offering competing
definitions of 'set' – each a replacement for the defective naïve conception.
But what of less dramatic cases, where there are no (known) contradictions
involved? How do we decide on the right definition to work with when
there is more than one live option?

In his famous book *Proofs and Refutations*, the Hungarian-born philosopher
of science Imre Lakatos (1922–74) argued that there is more to mathematics than merely proving theorems. We need to decide which mathematical
statements to try to prove, we need to agree upon the definitions of
the concepts employed in the statements in question, and we need to
know how to deal with alleged counterexamples.[15] Lakatos presents his

---

[15] Lakatos first presented this work in his Cambridge Ph.D. thesis in 1961 and in a series
of articles in the early 1960s. It eventually appeared posthumously in book form as
Proofs and Refutations in 1976.

case via a very entertaining dialogue about the Euler characteristic for polynomials.

The Euler characteristic, named for the Swiss mathematician Leonhard Euler (1707–83), describes the shape of a given space. However, the concept was originally applied to polyhedra – three-dimensional geometric solids with a number of flat faces and edges. The Euler characteristic χ is defined to be the number of vertices (V) of the polyhedra in question, minus the number of edges (E), plus the number of faces (F): $\chi = V - E + F$. Consider a couple of familiar polyhedra. A cube has 8 vertices, 12 edges, and 6 faces, so its Euler characteristic is 2. An octahedron has 6 vertices, 12 edges, and 8 faces, so its Euler characteristic is also 2. After considering other familiar examples, it is tempting to hypothesise that $\chi = 2$ for all polyhedra. This is known as *Euler's formula*: $\chi = V - E + F = 2$.[16]

One of the things that Lakatos brings out so nicely in *Proofs and Refutations* is that we cannot prove or disprove Euler's formula for polyhedra without simultaneously negotiating the definition of 'a polyhedron' and, in some cases, the definitions of 'edge', 'face', and 'vertex'. The reason for this is that there are counterexamples to Euler's formula, but the counterexamples are somewhat odd. Indeed, so odd are some of the counterexamples that it is tempting to decline to classify the objects in question as polyhedra, even though they satisfy the tentative definition of 'a polyhedron' we started out with. For example, consider the solid body bounded by two nested cubes (which do not touch at any point). This body apparently has 16 vertices (8 on the outside and 8 on the inside), 24 edges (12 on the outside and 12 on the inside), and 12 faces (6 on the outside and 6 on the inside), giving it a Euler characteristic of 4. Do we see this as a counterexample to Euler's formula or do we see it as a non-polyhedron? It depends on how serious we are about wanting Euler's formula to hold. We can make the notion of polyhedron restrictive so that Euler's formula does hold, or we can be more liberal about what we count as a polyhedron and then Euler's formula has counterexamples.[17] It turns out that there is a class of three-dimensional

---

[16] Not to be confused with another formula of complex analysis that also goes by this name: $e^{ix} = \cos x + i \sin x$.

[17] I'm just presenting the basic idea here. There are many interesting twists and turns, not to mention controversial conclusions, to be found in Lakatos's very rich book. I highly recommend reading it in its entirety.

figures for which Euler's formula holds, namely, convex polyhedra. And other polyhedra can have $\chi \neq 2$. But we might have just dug our heels in and insisted that for all polynomials $\chi = 2$, and explain away the counterexamples as pathological cases (or 'monster-barring', as Lakatos called the ad hoc dismissal of pathological cases).

Whether Lakatos's story about Euler's formula is historically accurate is debated, but the logical point about the relationship between definitions of mathematical concepts and theorems involving them is surely correct. Indeed, we can construct other examples. We could 'monster-bar' the Russell set and its like – refusing to accept them as sets – thus saving naïve set theory. We encountered the Weierstrass function in section 8.2.1: the function that's continuous on an interval but not differentiable anywhere in the interval. This function prompted debate over the definition of 'mathematical function'. In particular, the definition needed to determine whether the Weierstrass function and its kind were functions or whether they were to be dismissed as pathological cases.[18]

Monster-barring sounds ad hoc and thus unreasonable, but this is not always the case. In all the examples we've considered thus far, the more general definition has won out, and monster-barring, although a live option, was not invoked. But consider the following 'monster' – the Dirac delta function $\delta : \mathbb{R} \to \mathbb{R}$:

$$\delta(x) = 0, \quad \forall x \neq 0, \text{ and } \int_{-\infty}^{+\infty} \delta(x)\, dx = 1.$$

This 'function' can be thought of as having an infinite spike at the origin, with one unit of area under the spike.

To get the idea, consider a related family of functions $f_\epsilon : \mathbb{R} \to \mathbb{R}$:

$$f_\epsilon(x) = \begin{cases} 0, & x \notin (-\epsilon, \epsilon) \\ x/\epsilon^2 + 1/\epsilon, & 0 > x > -\epsilon \\ -x/\epsilon^2 + 1/\epsilon, & \epsilon > x \geq 0. \end{cases}$$

These piecemeal functions are zero everywhere except in a small region around the origin. Here they take the form of straight lines up to the point $(0, 1/\epsilon)$ and back down to zero again. They have spikes around the origin,

---

[18] See also the discussion of the Fundamental Theorem of Algebra in section 9.2.9 for another related case.

and these spikes are precisely high enough to ensure that the area under the curve in each case is one unit: $\int_{-\infty}^{+\infty} f(x)\, dx = 1$. Obviously, the smaller ϵ is, the higher the spike. Now consider the limit of this family of functions as ϵ approaches zero. That's the Dirac delta function! (It is named for the English physicist Paul A. M. Dirac (1902–84), who made fundamental contributions to quantum mechanics.)

The Dirac delta function proved to be very useful in physics. Indeed, Dirac's original proposal of the delta function was motivated by applications in quantum mechanics. But despite being useful in applications there is something very fishy about it. It's not exactly inconsistent, but it feels dangerously close. We normally think that in order for there to be area under a curve, there must be some (non-trivial) interval of the domain where the function is non-zero. But the Dirac delta function has its area concentrated at a point – the point where the spike is. The Dirac delta function was too weird to live, and too useful to die.[19] It was monster-barred as a function (hence the scare quotes earlier) but was rehabilitated as a different kind of mathematical object: a *distribution*. Distributions are generalisations of functions, and the Dirac delta function prompted the development of the theory of such objects.

So monster-barring, while apparently ad hoc in some cases, can be justified and is sometimes employed as a strategy. At the very least it is an option; Lakatos is right about that. And Lakatos is also right that *defining* mathematical concepts and *using* those concepts is a kind of juggling act. We start out with some intuitive mathematical concepts, derive some results using these concepts, refine our definitions in light of the results, and revisit the derivations. Neither definition nor derivation is logically or temporally prior to the other. This is in stark contrast to the received view of mathematics being a deductive science, cranking out theorems by agreed-upon rules and using well-defined concepts given in advance.

Discussion questions

1. Think of further examples where a good choice of mathematical notation can be thought to facilitate new mathematics.

---

[19] This phrase is borrowed and adapted from the 1998 Terry Gilliam film of Hunter S. Thompson's *Fear and Loathing in Las Vegas*.

2. Make a Möbius strip and verify that it is non-orientable. Cut along the midline of the strip. Do the same again. (Hint: make the original strip wide enough to accommodate the two cuts described here.)

3. Verify that a Klein bottle is a three-dimensional object with no inside or outside.

4. Verify that $\chi = 2$ for a tetrahedron (4 faces) and dodecahedron (12 faces).

5. Can you think of other counterexamples to Euler's formula?

6. Explain how the Dirac delta function can be thought of as the derivative of a discontinuous step function such as:

$$f(x) = \begin{cases} 0, & 0 > x \\ 1, & x \geq 0. \end{cases}$$

7. In empirical science, Quine and Duhem have argued that there is under-determination of theory by evidence, so that no single hypothesis is ever refuted by evidence. Rather, 'our statements about the external world face the tribunal of sense experience not individually but only as a corporate body' (Quine 1953, p. 41). So, according to Quine and Duhem (and Lakatos), in the empirical realm we can save a favourite hypothesis that's seemingly in conflict with empirical evidence, by shifting the blame to auxiliary hypotheses. Now reflect on the methodology of *Proofs and Refutations*, with its simultaneous defining concepts and deriving results about the concepts in question. Explain how this might be thought of as the mathematical equivalent of the Quine–Duhem–Lakatos view about confirmation and disconfirmation in empirical science. When seen in this light, Lakatos can be thought to be arguing for a kind of fallibilism about mathematics. Spell out the fallibilist view Lakatos is pushing for and note the points of contact with empirical science.

8. Could the Lakatos thesis about revising and reformulating definitions in the light of counterexamples be taken a step further, allowing logical notions such as 'proof', 'derivable', 'logical consequence', and 'validity' to be revised when a counterexample arises? Can you think of any examples where revising the logic might be seen as a rational response to a counterexample or a problem case?

Recommended further reading

There is very little in the philosophical literature on mathematical notation. Chapter 6 of Brown (2008) is essential reading in this regard. De Cruz and De Smedt (forthcoming), Munstersbjorn (1999), and Weber (forthcoming) are all relevant. There is quite a bit written on the mathematical side. See, for example, Davis and Hersh (1981). For the history of mathematical notation, see Cajori (1993). Related material can be found in discussions of visual/geometric methods versus algebraic methods, such as in Brown (2008, Chapter 3) and Giaquinto (2007). For more on non-deductive methods in mathematics, see Baker (2009c) and, of course, the classic Lakatos piece on mathematical definitions is his (1976). Other than looking at these few pieces, I'd recommend dipping into some of the relevant mathematics to get a feel for mathematical notation 'in the wild'. For example, Bold (1982), Massey (1989), and Spivak (2006) go into more detail on some of the examples discussed in this chapter.

Baker, A. 2009c. 'Non-Deductive Methods in Mathematics', in E. N. Zalta (ed.), *The Stanford Encyclopedia of Philosophy* (Fall 2009 edn), http://plato.stanford.edu/archives/fall2009/entries/mathematics-nondeductive/.

Bold, B. 1982. 'The Problem of Squaring the Circle', in *Famous Problems of Geometry and How to Solve Them*, New York: Dover, pp. 39–48.

Brown, J. R. 2008. *The Philosophy of Mathematics: A Contemporary Introduction to the World of Proofs and Pictures*, 2nd edn, London: Routledge.

Cajori, F. 1993. *A History of Mathematical Notation*, New York: Dover Reprints. (First published in two volumes by Open Court, London, 1929.)

Davis, P. J. and Hersch, R. 1981. *The Mathematical Experience*, Boston: Berkhäser.

De Cruz, H. and De Smedt, J. Forthcoming. 'Mathematical Symbols as Epistemic Actions', *Synthese*.

Giaquinto, M. 2007. *Visual Thinking in Mathematics*, Oxford University Press.

Lakatos, I. 1976. *Proofs and Refutations: The Logic of Mathematical Discovery*, Cambridge University Press.

Massey, W. S. 1989. *Algebraic Topology: An Introduction*, New York: Springer-Verlag.

Munstersbjorn, M. M. 1999. 'Naturalism, Notation, and the Metaphysics of Mathematics', *Philosophia Mathematica*, 7(2): 178–99.

Spivak, M. 2006. *Calculus*, 3rd edn, Cambridge University Press.

Weber, Z. Forthcoming. 'Figures, Formulae, and Functors', in S. Shin and A. Moktefi (eds.), *Visual Reasoning with Diagrams*, Springer.

9   Epilogue: desert island theorems

Beauty is the first test: there is no permanent place in the world for ugly mathematics.

G. H. Hardy (1877–1947)[1]

You know the old question about which 20 books, 20 albums, 20 movies, or whatever you'd like to have with you if you were stranded on a desert island? Well, in this chapter I'll give you my top 20 mathematical theorems for desert island-bound philosophers. We look at a number of mathematical results that have some philosophical interest, or in some cases are just very cool pieces of mathematics. (Alternatively, you might think of this chapter as 20 theorems you should come to terms with before you die.) Of course, this is just my top 20 theorems. If you don't like my choices, feel free to construct your own list. For good measure I throw in a few famous open problems and interesting numbers to round out my desert-island survival kit.

9.1 Philosophers' favourites

The theorems in this section are well known by philosophers and rightly get a great deal of attention in philosophical circles. These are the obvious choices for desert island theorems, but in some cases you'd be disappointed to be stuck with just these. You wouldn't be disappointed because they are uninteresting or trivial; you'd be disappointed because they are just a bit too obvious. Everybody would have these![2] In any case, the theorems

---

[1] G. H. Hardy, *A Mathematician's Apology*, Cambridge University Press, 1940 [1967], p. 14 [p. 85].

[2] Just as if you went to the desert island with only cinema classics such as *Casablanca*, *Citizen Kane*, *Vertigo*, *The Godfather*, and so on, you'd eventually crave something a bit less

below are the classics – the obvious ones that almost anyone would put high on their list. (These are the *Citizen Kane*s and *Vertigo*s of the maths world.)

9.1.1 Tarski–Banach Theorem (1924)

The Tarski–Banach Theorem is a theorem of topology named after the Polish logician and mathematician Alfred Tarski (1901–83) and the Polish mathematician Stefan Banach (1892–1945), and was first published in a paper appearing in 1924. The theorem states that a solid sphere can be decomposed into a finite number of pieces, the pieces moved around via rigid rotations and translations, and recombined into two spheres, each equal in volume to the first. The theorem crucially depends on the axiom of choice, which allows for the decomposition in question into non-measurable sets. It is deeply counterintuitive, so much so that it is seen by some as a *reductio* of the axiom of choice.

9.1.2 Löwenheim–Skolem Theorem (1922)

If a system of first-order sentences has a model at all, it has a countable model. This theorem was discussed at length in section 2.1.

9.1.3 Gödel's Incompleteness Theorems (1931)

I know, I'm cheating here by including two theorems under one heading, but these two are very closely related – they're the mathematical equivalent of a two-part movie (such as *Kill Bill*). Gödel's First Incompleteness Theorem states that any consistent system of axioms of sufficient complexity to be of mathematical interest will have true sentences that are not derivable within the system. Gödel's Second Incompleteness Theorem states that any consistent system of axioms of sufficient complexity to be of mathematical interest cannot prove its own consistency. These theorems were discussed at length in section 2.2.1.

familiar and perhaps a bit offbeat, such as *The Good, the Bad and the Ugly*, *The Big Lebowski*, *Stranger than Paradise*, or *12 Monkeys*.

9.1.4 Cantor's Theorem (1891)

For any set A, $|\mathcal{P}(A)| > |A|$. We discussed and proved this theorem in section 1.

9.1.5 Independence of continuum hypothesis (1963)

The continuum hypothesis, $2^{\aleph_0} = \aleph_1$, is independent of standard Zermelo–Fraenkel set theory with the axiom of choice (ZFC), in the sense that the continuum hypothesis is consistent with ZFC, but so is the negation of the continuum hypothesis. This result was discussed in further detail in section 2.3.1

9.1.6 Four-Colour Theorem (1976)

The Four-Colour Theorem was the first major mathematical result to rely on computers for the proof. The theorem states that four colours are all that are needed to colour a map in such a way that each country is distinguishable by colour from those with which it shares a border. After many years of attempts, the theorem was finally proved in 1976 by Kenneth Appel (1932–) and Wolfgang Haken (1928–). The philosophical interest of the theorem revolves around the question of what counts as a mathematical proof. This proof was computer-assisted, in the sense that large portions of the proof were carried out by a computer without any human mathematician checking the proof in its entirety. The proof of the theorem raises questions about whether a 'proof' that no one understands is a proof, and about the relative reliability of computer methods compared to the methods of human mathematicians, to name but a couple of such issues. Computer proofs are more common these days and do not generate quite the controversy this one did. But you would expect the first such proof to prompt healthy debate on the nature of mathematical proof, and this one certainly did that.

9.1.7 Fermat's Last Theorem (1995)

This theorem has a rich and colourful history, starting with a note by the French mathematician Pierre de Fermat (1601–65) in 1637 in his copy of the margin of the ancient Greek text *Arithmetica* by Diophantus. Famously, Fermat wrote:

> It is impossible for a cube to be the sum of two cubes, a fourth power to be the sum of two fourth powers, or in general for any number that is a power greater than the second to be the sum of two like powers. I have discovered a truly marvellous demonstration of this proposition that this margin is too narrow to contain.

What Fermat claims here is that he has a proof of the theorem that there are no non-zero solutions to the equation $x^n + y^n = z^n$, where x, y, z, and n are integers, for $n > 2$.

Part of the attraction of the theorem is that it is so simple to state and, according to Fermat, it has a 'marvellous' and, presumably, fairly elementary proof. It now seems that Fermat was mistaken about having a proof. For over 350 years, attempts were made, hopes raised, and flaws in alleged proofs found, until it was finally laid to rest in 1995 by the English mathematician Andrew Wiles, after initially putting forward an incomplete proof in 1993.[3] The proof ran to over a hundred pages in length (much too long for the margin of Fermat's copy of *Arithmetica*!) and employed mathematical methods that went way beyond anything available at Fermat's time. Indeed, from a mathematical point of view, the value of the theorem is more in the work it has spawned (such as Wiles's and others' work on elliptic curves and modular forms), rather than in the theorem itself. Still, with such a history, Fermat's Last Theorem is one of the few mathematical results to have captured the public imagination and entered into popular culture. The philosophical interest in the theorem has largely dried up since it was finally proved. Before that, it was regularly used, along with Goldbach's conjecture (see below), as an example of a well-known open mathematical question. Thanks to Andrew Wiles, philosophers now have to look for other examples of open mathematical questions and unproven conjectures. Fortunately, there is no shortage of these, but few are as simply stated and as easily understood as Fermat's Last Theorem.

---

[3] The gaps in the 1993 proof were finally closed late in 1994 by Wiles and Richard Taylor, and published in 1995.

9.1.8 Bayes's Theorem (1763)

Named in honour of the Reverend Thomas Bayes (1702–61), this theorem of probability theory is a candidate for the mathematical theorem with the largest number of philosophical applications and the greatest philosophical interest. It also serves as a nice illustration of how rather elementary mathematics (as this is) can be philosophically very rich. On the face of it, the theorem simply tells us how to calculate a conditional probability, $P(A|B)$ – the probability of A, given B.

In its simplest form this theorem states:

$$P(A|B) = \frac{P(B|A)P(A)}{P(B)}$$

for $P(B) \neq 0$. (This form of the theorem is sometimes called 'the Inverse Probability Law'.) A somewhat more useful version of the theorem is

$$P(A|B) = \frac{P(B|A)P(A)}{P(B|A)P(A) + P(A^c)P(B|A^c)}$$

for $P(B) \neq 0$ and where A^c is the complement of A. The general form of the theorem, for an arbitrary partition A_i of the event space in question, is:

$$P(A_j|B) = \frac{P(B|A_j)P(A_j)}{\sum_i P(B|A_i)P(A_i)}$$

for $P(B) \neq 0$. The proof is straightforward. The simplest version above follows almost immediately from the definitions of the relevant conditional probabilities: $P(A|B) = P(A \cap B)/P(B)$ and $P(B|A) = P(A \cap B)/P(A)$. The final version is an immediate consequence of the Law of Total Probability $P(B) = \sum_i P(B|A_i)P(A_i)$, with the second version a special case of the final version.

One of the most significant applications of this theorem is in belief updating. If we think of probabilities as measures of our degrees of belief, Bayes's Theorem tells us how to update those beliefs in the light of new evidence. A particularly important example here is the case of scientific hypothesis testing. Consider your initial belief $P(H)$ (or *prior probability*) in some scientific hypothesis, H. Bayes's Theorem tells you how to revise that belief to form your *posterior probability* $P(H|E)$, once you learn about the evidence E. Despite its intuitive appeal, this, application of the theorem is rather controversial and lies at the heart of a major dispute in statistics and philosophy of statistics. Some suggest that it does not make sense to talk of probabilities prior to evidence, as the application just mentioned requires. Others see nothing wrong with this. The former group typically understand probabilities as ratios of frequencies, and this prohibits them from using Bayes's Theorem in cases where the prior probabilities lack an objective (frequentist) statistical basis. Those who see nothing wrong with probabilities prior to evidence deny the frequentist interpretation of probability theory and opt for a subjectivist account, according to which probabilities are degrees of belief. This subjectivist camp find that they can use Bayes's Theorem in many cases where frequentists cannot, and this is the main reason that the subjectivist camp are somewhat misleadingly referred to as 'Bayesians'. It's a misleading name because Bayes's Theorem is accepted by all parties (it is, after all, a theorem). Nevertheless, Bayes's Theorem is at the centre of this major (and often *heated*) controversy over the interpretation of probability theory and the correct methodology in statistics.

9.1.9 Irrationality of $\sqrt{2}$ (*c.* 500 BCE)

The Pythagoreans were a group of philosophers (cum mystical religious cult) centred on the great Greek philosopher and mathematician Pythagoras (*c.* 570–*c.* 490 BCE). Although Pythagoras is most famous for the theorem named after him ('the sum of squares of the sides of a right-angled triangle is equal to the square of the hypotenuse'), it is unlikely that he in fact proved that particular theorem. The Pythagorean world-view revolved around mathematics to such an extent that it was believed that the world itself was number. It's not clear exactly what this was supposed to mean, but by 'number' they meant integers and ratios of integers. So they were committed to the view that the world was (or at least was describable in terms of) rational numbers. It thus came as a bombshell when one of their own showed that irrational magnitudes, such as $\sqrt{2}$, existed in nature. There is a suggestion that there was a great scandal when the existence of irrational magnitudes was leaked beyond the Pythagoreans, with one version of the story suggesting that when the source of the leak was found, the person in question was drowned at sea as punishment.

So much for the myths and legends, let's get back to the mathematics. One proof of the theorem that $\sqrt{2}$ is irrational provides a nice example of a *reductio ad absurdum* proof and is worth rehearsing here. Assume, by way of contradiction, that $\sqrt{2}$ is *rational*. We can thus write $\sqrt{2} = p/q$ where p and q are integers with no common divisors. But this implies that $2q^2 = p^2$. So p^2 is even and hence p is even. If p is even, we can write it as $p = 2r$ for some integer r and we thus have from above that $q^2 = 2r^2$ and hence q is also even. This is a contradiction, since we assumed that p and q had no common divisors. Therefore $\sqrt{2}$ is irrational.

9.1.10 Infinitude of the primes (*c.* 300 BCE)

Euclid (*c.* 300 BCE) proved that there are infinitely many prime numbers. This theorem was proved and discussed in section 2.

9.2 The under-appreciated classics

Now for some very important mathematical theorems that don't get as much attention in philosophical circles as perhaps they deserve. Still, these theorems are ones that every philosopher of mathematics would be well advised to be familiar with. (These are the *Big Lebowskis* and *Stranger than Paradises* of the maths world.)

9.2.1 Borsuk–Ulam Theorem (1933)

This theorem of algebraic topology is named in honour of the Polish-US mathematician Stanisław Ulam (1909–84) and the Polish mathematician Karol Borsuk (1905–82). Let $f : S^n \to R^n$ be a continuous map, then there exists an $x \in S^n$ such that $f(x) = f(-x)$. In the special case of this theorem where $n = 2$, we have f as a map from the surface of a sphere to R^2. The usual corollary used to illustrate the theorem is that on the assumption that the earth is topologically equivalent to a sphere[4] and that temperature and atmospheric pressure vary continuously across the earth's surface, there are two antipodal points on the earth's surface with precisely the same temperature and atmospheric pressure.

---

[4] That is, we need to assume that the earth does not have any tunnels. Of course, the earth does have tunnels but we can avoid this assumption by taking a tunnel-less surface a metre above the actual surface of the earth.

What is philosophically interesting about this theorem is that it is a piece of pure mathematics, but it seems to tell us how things must be in the physical world. Indeed, were we to find two such antipodal points with the same temperature and atmospheric pressure, and were puzzled by their existence, the theorem would offer some explanation: algebraic topology guarantees that there must be two such points. In a sense, the presence of two such points has nothing to do with meteorology and everything to do with topology. Elsewhere I have leaned on this theorem in order to argue for the existence of mathematical explanations of physical phenomena.

This theorem is closely related to a number of other interesting theorems, most notably Brouwer's Fixed-Point Theorem and the Ham Sandwich Theorem (and its two-dimensional counterpart, the Pancake Theorem). Brouwer's Fixed-Point Theorem states that any continuous map from an n-ball to an n-ball has a fixed point (i.e., a point mapped to itself). The Ham Sandwich Theorem tells us that any n finite measurable regions of R^n (e.g., in R^3, we could have a piece of ham and two pieces of bread), can be each cut in half (by volume) by a single $n-1$ hyperplane (in R^3, we can cut each component of the ham sandwich in half with a single cut with a plane).

9.2.2 Riemann Rearrangement Theorem (1854)

This theorem is named in honour of the German mathematician Bernhard Riemann (1826–66). A conditionally convergent series $\sum_{j=1}^{\infty} a_j$ is one that is convergent, but the related series $\sum_{j=1}^{\infty} |a_j|$ is divergent. A series is absolutely convergent if both of these series converge. The Riemann Rearrangement Theorem states that the terms in a conditionally convergent series can be rearranged to converge to any value or to diverge to plus or minus infinity. For example, the series

$$\sum_{j=1}^{\infty} \frac{1}{2^j} (-1)^{j-1} \frac{2^j}{j} = \sum_{j=1}^{\infty} \frac{(-1)^{j-1}}{j}, \tag{9.1}$$

converges (to ln.2), but the related series (the harmonic series)

$$\sum_{j=1}^{\infty} \left| \frac{(-1)^{j-1}}{j} \right|$$

is divergent. This means that the terms in the first series can be rearranged to yield any value as the value of the sum. That is, the value of the sum ln .2 in equation 9.1 depends crucially on the order of the terms. This is very counterintuitive. In finite cases, the value of a sum does not depend on the order of the terms. The result turns, in part, on the way 'summation' is defined in the infinite case (in terms of the limit of partial sums) and in part on the oddity of conditionally convergent series themselves. The latter are unstable, temperamental creatures.

Apart from being an interesting mathematical result in its own right, the Riemann Rearrangement Theorem has recently been put to good philosophical use in generating a paradox in decision theory: the Pasadena paradox. The Pasadena game consists in a sequence of tosses of a fair coin until the first head appears. At the appearance of the first head, the game is over. The pay-off schedule for the game is given by:

> If the first head appears on toss n, the pay-off is given by $\$(-1)^{n-1}2^n/n$, where a negative amount indicates the punter pays the bookie and a positive amount indicate that the bookie pays the punter.

What's interesting about this game is that the expected utility calculation involves a conditionally convergent series. Indeed, it involves precisely the conditionally convergent series we considered in equation 9.1 above. The Riemann Rearrangement Theorem is then invoked to show that the resulting expected utility can be made to converge to any finite value or diverge to positive or negative infinity – it all depends on the order of the terms in the series. But the terms in the series do not come in any natural order, and therein lies the problem. The value of the game crucially depends on something that is neither specified by the game nor considered part of the usual specification for a well-posed decision problem. The natural conclusion is that, despite looking like a well-posed problem of decision theory, it is in some sense defective, in that it does not have an expectation.[5]

9.2.3 Gauss's Theorema Egregium (1828)

This theorem of differential geometry is a candidate for my all-time favourite mathematical theorem. This theorem was proven by Carl Friedrich Gauss and its name is Latin for 'remarkable theorem', so it seems that this theorem might also have made Gauss's desert island list. The

---

[5] See Colyvan (2006) and Nover and Hájek (2004) for more on the Pasadena paradox.

theorem states that the curvature of a manifold can be specified locally: the curvature of a surface can be determined entirely by measuring angles and distances on the surface. It is natural to think of a curved surface as being curved in another dimension. For example, we have the two-dimensional surface of the earth curved in three-dimensional space. But Gauss's Theorema Egregium tells us that there is no need to think of curvature from this extrinsic point of view, with curvature through another dimension. We can make perfect sense of an n-dimensional surface being intrinsically curved – without an $(n + 1)$-dimensional space in which the surface in question resides. This result lies at the heart of differential geometry and has important applications in field theories in physics. It allows us, for example, to make sense of the curved four-dimensional space-time manifold in general relativity without the need for a fifth dimension in which the curvature occurs.

9.2.4 Residue Theorem (1831)

This result is due to the French mathematician Augustin-Louis Cauchy and is one of the jewels of complex analysis. The theorem concerns integrals of holomorphic functions around closed contours in the complex plane.[6] The theorem states that the line integral of a holomorphic function $f(z)$ around a closed curve can be calculated in terms of the residues of poles of the function:

$$\oint_\gamma f(z)dz = 2\pi i \sum_{a\in A} \text{Res}_{z=a_i} f(z),$$

where A is the set of poles of the function enclosed within the (oriented) contour γ and $\text{Res}_{z=a_i} f(z)$ is the residue of $f(z)$ at $z = a_i$. The poles of a complex-valued function are singularities where $f(z)$ approaches infinity as z approaches the point in question. For example, $f(z) = 1/z$ has a single pole at $z = 0$. The residue of a complex function at an isolated pole is a complex number and can be defined in terms of the Laurent series expansion of the function around the point in question (the details needn't concern us here). The upshot is that the contour integral of a complex function depends only on what happens at a limited number of singularities

---

[6] Holomorphic functions are complex functions that are differentiable.

inside the contour. So although a line integral looks as though it is local (i.e., depends on the behaviour of the function along the contour in question), in fact it is determined by non-local features of the function (i.e., what happens at singularities remote from the contour in question). This is the mathematical equivalent of action at a distance.

The theorem is crucial in calculating many integrals, including many real-valued ones. The later applications of this theorem are philosophically interesting because they involve finding an answer to a problem in real analysis via an essential excursion into complex analysis. This raises questions about intrinsic explanation in mathematics. You'd expect to be able to calculate real integrals via real methods and, moreover, any mathematical explanations of why the integrals take the values they do should be made in terms of real analysis. The Residue Theorem and its applications to real analysis raise doubts about both these expectations. Sometimes the only means we have of calculating real integrals is via the Residue Theorem and the associated excursion into the complex domain. Moreover, it can be argued that this excursion is more than just a useful calculational trick: in at least some cases, the *explanation* of the answer is given by the Residue Theorem (or so it seems to me). This, in turn, suggests that intrinsic explanations (in this case, real-analysis explanations of real-analysis facts) are not always possible. Alternatively, we might draw the conclusion that the explanations in question via the Residue Theorem are, despite appearances, intrinsic; it's just that real analysis and complex analysis are more closely connected than we might initially think. Such applications of the Residue Theorem to real analysis thus raise interesting questions about the boundaries between the various branches of mathematics and force us to ponder intra-mathematical explanations.

9.2.5 Poincaré conjecture (2002)

This theorem was originally conjectured by the French mathematician and physicist Henri Poincaré but has since been proved, and thus has been promoted to the status of theorem. It is still known, however, by its old 'conjecture' title. The theorem states that any simply connected, closed three-dimensional manifold is topologically equivalent to a 3-sphere. A 3-sphere is the three-dimensional generalisation of the more familiar (two-dimensional) sphere – it is a three-dimensional surface consisting of the set of points some specified distance from a designated point (the centre) in four-dimensional Euclidian space. A simply connected manifold is one that is path-connected (i.e., there is a continuous path from any point on the manifold to any other point in the manifold) and such that every closed curve in the manifold can be continuously shrunk to a point. The conjecture is thus about a space that is locally like three-dimensional space but is finite in volume, without boundary, and has the property that any closed curve can be continuously tightened to a point. Any such space is topologically equivalent to a 3-sphere, which means that either the space in question is a 3-sphere or it can be continuously deformed into a 3-sphere. The conjecture thus provides a topological characterisation of a 3-sphere.

The conjecture was one of the most outstanding problems in mathematics before it was finally proved by Russian mathematician Grigori Perelman (1966–) in 2002. Perelman was awarded the Fields medal – the most prestigious prize in mathematics – for his proof, but he did not accept the medal. The Poincaré conjecture was also one of seven outstanding mathematics problems identified in 2000 by the Clay Mathematics Institute. The seven problems in question are known as *the Millennium Problems* and each attracts a prize of 1 million US dollars for a correct solution. So far the Poincaré conjecture is the only one of the Millennium Problems to have been solved. In 2010 the Clay Mathematics Institute determined that Perelman had indeed proved the Poincaré conjecture and had thus met the conditions of the prize. Perelman, however, turned down the prize, claiming that his contribution towards the proof was no more significant than the contributions of others who went before him.

9.2.6 Prime Number Theorem (1849)

There are a lot of prime numbers in the lower reaches of the natural numbers, but they seem to thin out as we proceed into the higher reaches of the natural numbers. For example, between 1 and 20 there are eight primes: 2, 3, 5, 7, 11, 13, 17, and 19. Between 101 and 120 there are only five of them: 101, 103, 107, 109, 113; between 1,001 and 1,020 there are only three: 1,009, 1,013, 1,019; and between 10,001 and 10,020 there are two: 10,007 and 10,009. This leads to a very interesting question about the number of prime numbers $\pi(n)$ less than some given number, n. After conjectures about this prime-counting function by Adrien-Marie

Legendre (1752–1833) and a young (15-year-old) Gauss, the result known as the Prime Number Theorem was first proposed by Gauss in 1849 and proved in 1896 independently by the French mathematician Jacques Hadamard (1865–1963) and the Belgian mathematician Charles-Jean de la Vallée Poussin (1866–1962). The theorem states that $\pi(n) \sim \int_2^n dx/\ln x$. There are interesting further questions about the distribution of the primes. The Riemann zeta function and the Riemann hypothesis about this function are closely tied to the distribution of the primes (see the section on open problems below).

9.2.7 The Fundamental Theorems of Calculus (*c*. 1675)

I know, I'm cheating again – sneaking two theorems in under one heading, but again these two are very closely related. These two are perhaps not so interesting philosophically – at least I can't think of any particular philosophical issues raised by them – but they are such important results that they earn their place in this list for their mathematical interest alone. The first theorem forges a link between the algebraic indefinite integral and the geometric definite integral. The theorem states that if f is continuous on a closed interval $[a, b]$ and F is the antiderivative of f on $[a, b]$, then $\int_a^b f(x)\, dx = F(b) - F(a)$.

The second theorem tells us that differentiation and integration are two sides of the one coin. In advance, there's no reason to expect this. After all, differentiation arose from the problem of finding the tangent to a curve at a given point, with the physical application of making sense of instantaneous velocities and accelerations. Integration arose from finding the area under a curve, with the physical application of calculating the work done on a body being moved with a varying force. The theorem tells us that if f is a real-valued function, continuous on an open interval (a, b), with c any point in (a, b) and $F(x) = \int_c^x f(t)\, dt$, $F'(x) = f(x)$ (where $F'(x)$ is the derivative of $F(x)$).

These cornerstones of calculus are perhaps best thought of as collaborative efforts by a number of mathematicians over a long period of time. The first general statement and proof of the theorems was by the English mathematician Isaac Barrow (1630–77), with the development of the associated mathematics by the English mathematician and physicist Isaac Newton (1643–1727) and the German philosopher and mathematician

Gottfried Leibniz (1646–1716). Subsequent developments involving the rigourisation of the calculus were made by many, most notably by Augustin-Louis Cauchy.

9.2.8 Lindemann's Theorem (1882)

As we saw in Chapter 8, this theorem states that π is transcendental and has as a corollary that it is impossible to square the circle with straightedge and compass. There's an interesting aside on Lindemann's Theorem. There was a bizarre incident in Indiana in 1897 when an amateur mathematician was convinced that he had squared the circle and tried to get a bill passed in the Indiana State Legislature: 'A Bill for an act introducing a new mathematical truth and offered as a contribution to education to be used only by the State of Indiana free of cost by paying any royalties whatever on the same, provided it is accepted and adopted by the official action of the Legislature of 1897.' The bill was almost passed but for a last-minute intervention from a Purdue University mathematics professor.[7] There is also the famous line in the Bible (1 Kings 7:23) that seems to imply that $\pi = 3$. So it might be argued that another corollary of Lindemann's Theorem is that the Bible contains at least one false claim.

9.2.9 Fundamental Theorem of Algebra (1816)

This theorem is usually attributed to Carl Friedrich Gauss, but the originator of the geometric interpretation of complex numbers, Jean-Robert Argand (1768–1822), also has a legitimate claim to it.[8] The theorem is as follows: a polynomial of degree n has n roots (with some of them possibly degenerate).

---

[7] The Indiana construction seemed to imply two values for π: 3.2 and 4. Perhaps they should also have considered repealing the law of non-contradiction!

[8] Like many of the older theorems in this chapter, it is not straightforward to say when they were proved and even who proved them. Many of the theorems had several proofs advanced by a number of different mathematicians over many years. Many of these proofs had gaps in them, at least by today's standards. Sometimes the theorem in question was attributed to the first person to provide a rigorous gapless proof, at other times the theorem is attributed to the first person to have had a decent stab at the proof. The Fundamental Theorem of Algebra has a rather interesting history, with quite a few prominent figures involved in the search for a rigorous proof. In the end it comes down to Gauss and Argand as the main contenders, though.

The roots of a polynomial are the values of the variables such that the value of the polynomial is zero. For example, the second-degree polynomial, $x^2 - 1 = (x - 1)(x + 1)$ has roots $+1$ and -1, and the third-degree polynomial $x^3 + x^2 - 2x = x(x + 2)(x - 1)$ has three roots 0, -2, and 1. Polynomials such as $x^2 - 2x + 1 = (x - 1)(x - 1)$, which have only one root, are said to have two degenerate roots or, alternatively, two roots that coincide. The motivation for the latter claim comes from considering the factorisation of the polynomial in question – there are two terms in the factorisation but each of these terms is identical.

Next consider the polynomial $x^2 + 1$. This appears to have no roots. This is where things get interesting. $x^2 + 1$ has no *real* roots, but it still has two roots; they are the complex roots $+i$ and $-i$, where $i = \sqrt{-1}$. Indeed, the complex numbers are needed to make the Fundamental Theorem of Algebra hold, and the theorem thus provides a kind of validation for the complex numbers.[9] It's a nice example of justifying a piece of mathematics by its fruits. The fruit, here, being particularly juicy, namely, the Fundamental Theorem of Algebra.

The proofs of this theorem are also interesting. There are a number of proofs, all of which rely on quite a bit of mathematics outside of algebra. There are topological proofs, complex analysis proofs (involving analytic functions), and the closest thing to an algebraic proof uses some key results from calculus (such as the Mean-Value Theorem). As we saw before in the discussion of the Residue Theorem, if such detours into other branches of mathematics are essential, this raises interesting philosophical questions about the boundaries of the various branches of mathematics. Perhaps there is no real difference between algebra and complex analysis. Perhaps the connections between seemingly disparate branches of mathematics that are forged by theorems such as this are what mathematics is all about, or at least are what the most significant mathematics is about.

9.2.10 Fundamental Theorem of Arithmetic (*c.* 300 BCE)

The Fundamental Theorem of Arithmetic states that every natural number greater than 1 is expressible in exactly one way (up to rearrangement) as

---

[9] Historically, the theorem played an important role in the acceptance of the complex numbers, as did the various applications of complex analysis.

the product of one or more primes. For example, primes p are expressible as $p = p \times 1$, and some of the smaller composites are $4 = 2^2$, $6 = 2 \times 3$, $8 = 2^3$, $9 = 3^2$, $10 = 2 \times 5$, $12 = 2^2 \times 3$. Note that although 12 can be factored two different ways – as 6×2 or 3×4 – its prime factorisation is unique. The theorem tells us that in a very important sense, the prime numbers are the basic building blocks of number theory – every natural number has a unique prime factorisation or, if you prefer, has a unique prime signature or coding. This theorem is due to Euclid.

The Fundamental Theorem of Arithmetic has many interesting applications, one of which is in public-key cryptography. Cryptography is about coding and decoding messages. The more familiar codes allow one to freely encode and decode, when in possession of the key. But in various important applications, we require many users to be able to encode but only a few to decode.[10] It turns out that codes can be constructed in such a way that encoding relies on a large composite number, but the decoding relies on this number's unique prime factorisation (guaranteed by the Fundamental Theorem of Arithmetic). Such codes are easily constructed by multiplying together a string of large prime numbers. The crucial points here are (i) the result of such multiplications are unique composite numbers; (ii) multiplication is computationally easy; and (iii) recovering prime factorisations for a given (large) composite number is typically computationally hard.

As in any 'desert island' list, there will be unlucky items left out, despite having a very good case for inclusion. Some of those theorems I couldn't find room for here are: the Classification of Finite Simple Groups, the Squeezing Theorem, Arrow's Theorem, Brouwer's Fixed-Point Theorem, the Compactness Theorem, the Mean-Value Theorem, the Central Limit Theorem, L'Hôspital's rule, the Fan Theorem, the Intermediate-Value Theorem, the Gauss–Bonnet Theorem, Euler's equation, Euler's formula, the Unisolvence Theorem, and quite a few others. The interested student may like to seek out these theorems.[11]

---

[10] Such applications arise, for example, in banking and finance.

[11] I am grateful to Rachael Briggs, Alan Hájek, Chris Hitchcock, and Jack Justus for their suggestions for this chapter and for sharing their desert island theorems with me.

9.3 Some famous open problems

9.3.1 Riemann hypothesis

The Riemann hypothesis is arguably the most outstanding unsolved problem in mathematics. It was first articulated by Bernhard Riemann in an address to the Berlin Academy in 1859. The address was called 'On the Number of Prime Numbers Less Than a Given Quantity', and among the many interesting results and methods contained in that paper was Riemann's famous hypothesis: all non-trivial zeros of the zeta function, $\zeta(s) = \sum_{n=1}^{\infty} n^{-s}$, have real part 1/2. Although the zeta function as stated and considered as a real-valued function is defined only for $s > 1$, it can be suitably extended. It can, as a matter of fact, be extended to have as its domain all the complex numbers (numbers of the form $x + yi$, where x and y are real numbers and $i = \sqrt{-1}$) with the exception of $1+0i$ (at which point the zeta function is undefined). This extended zeta function takes the value zero for infinitely many complex numbers. For instance, all the negative even integers are zeros of the zeta function. These, however, are the *trivial* zeros. The Riemann hypothesis is thus the conjecture that all the other zeros (and there are also infinitely many of them) have the form $1/2 + yi$. This hypothesis is of crucial importance in analytic number theory. The zeta function is very closely related to the prime counting function $\pi(n)$ (which is the number of prime numbers less than or equal to some natural number n). Indeed, the zeta function 'encodes' important information about the distribution of primes, and the location of the non-trivial zeros of the zeta function is crucial in all of this. Again we see complex analysis playing an indispensable role in other branches of mathematics – this time it's number theory that requires complex analysis in apparently essential ways.[12]

9.3.2 The twin prime conjecture

Apart from 2, all other prime numbers are odd and some pairs of primes, such as $(5, 7)$, $(17, 19)$ and $(101, 103)$, are consecutive odd numbers. These

---

[12] The English mathematician G. H. Hardy, before embarking on a particularly dangerous journey early in the twentieth century, once sent a postcard (falsely) claiming to have solved the Riemann hypothesis. He did this as a kind of insurance. He figured that God wouldn't allow him to die with such undeserved glory attached to his name.

are known as *twin primes*. More precisely, twin primes are pairs of primes of the form $(p, p + 2)$. The conjecture that there are infinitely many such pairs is known as the *twin prime conjecture*. The conjecture is thought to be true, although, to date, there is no proof.

9.3.3 Goldbach's conjecture

This conjecture, named after the German mathematician Christian Goldbach (1690–1764), is that every even integer greater than 2 can be expressed as the sum of two primes. For example, we have $4 = 2 + 2$, $6 = 3 + 3, 8 = 5 + 3, 10 = 7 + 3$ (or $5 + 5$). It was first articulated in a letter from Goldbach to the great Swiss mathematician Leonhard Euler in 1742.

9.3.4 Infinitude of the Mersenne primes

A Mersenne prime is a prime number of the form $2^p - 1$, where p is a positive integer. They are named after the French mathematician Marin Mersenne (1588–1648). A few examples of Mersenne primes: $2^2 - 1 = 3$ is a Mersenne prime, as is $2^3 - 1 = 7$, $2^5 - 1 = 31$, and $2^7 - 1 = 127$. Notice that in all these cases, p is prime. This is no coincidence. It is fairly straightforward to show that if $2^p - 1$ is prime, then p is also prime. (The converse, however, is not true, and it is this that makes searching for Mersenne primes difficult. One open question associated with them is how many of them there are. Only a small number of Mersenne primes are known – at the time of writing, fewer than 50 – but for all we know, there could be infinitely many of them.)

9.3.5 Is there an odd perfect number?

A perfect number is a positive integer n that is the sum of its positive divisors, excluding n itself. The first perfect number is $6 = 1 + 2 + 3$. Others are 28, 496, and 8,128. Perfect numbers are closely tied to Mersenne primes by the following theorem: n is an even perfect number if and only if it has the form $2^{p-1}(2^p - 1)$, where p is prime and $2^p - 1$ is a Mersenne prime. All known perfect numbers are even, but it is a very old and famous open question as to whether there is an odd perfect number.

9.4 Some interesting numbers

- 0: zero is the additive identity. That is, for all numbers a, $a + 0 = a$. It is the multiplicative annihilator. That is, for all numbers a, $0 \cdot a = 0$. It is the cardinality of one of the strangest sets of all: the empty set, $\emptyset = \{x : x \neq x\}$. It is the exponent such that for all a, $a^0 = 1$. It is the only number that does not have a multiplicative inverse: for all numbers n, $n/0$ is undefined. Zero really is a special case of a natural number. Familiarity makes it easy to forget just how strange zero is. Consider the following proof that zero is not a number at all. If I say to you that I have played a number of games of first-grade Australian rules football for the Geelong football club, when in fact (sadly) I have never played any such games, then surely I've lied to you. Pointing out that the number of games in question is zero does not get me off the hook on the charge of lying. (Early concerns about extending the number system to include zero revolved around issues not too dissimilar from this somewhat tongue-in-cheek proof.)

- 1: the first natural number (or the first after zero, if you want to include zero as a natural number).[13] It is the multiplicative identity. That is, for all numbers a, $a \cdot 1 = a$.

- The Golden Ratio $\varphi = \frac{1+\sqrt{5}}{2} = 1.6180339\ldots$: this is the ratio (typically of lengths) such that for $a > b$, $a + b/a = a/b$. In some 'new age' circles this number is thought to be the key to the universe. It's not that, but it is pretty cool and pops up all over the place. It is closely related to the Fibonacci sequence of numbers, named after Leonardo of Pisa (aka Fibonacci) ($c.$ 1170–$c.$ 1250): 1, 1, 2, 3, 5, 8, 13, 21, 34, ..., where the first two are defined to be both 1, and thereafter the nth Fibonacci number F_n is the sum of F_{n-1} and F_{n-2}.[14] The Golden Ratio φ is the limit of the ratio of Fibonacci numbers: $\varphi = \lim_{n \to \infty} F_{n+1}/F_n$. Both the Golden Ratio

---

[13] It is not just 0 and 1 that are interesting; all the natural numbers are interesting. Suppose, by way of contradiction, that there are some uninteresting natural numbers. Now consider the smallest such number, b. The fact that b is the smallest uninteresting number surely makes it exceptional and therefore interesting. We thus have a contradiction, so all the natural numbers are interesting.

[14] Some presentations include 0 as the first Fibonacci number.

and Fibonacci sequence turn up in nature in many places, especially in various kinds of growth.[15]

- 2: the only even prime.

- e: this transcendental number is the base of the natural logarithm. It is the unique real number such that $d/dx\, e^x = e^x = \int_{-\infty}^{x} e^t\, dt$.

- π: this transcendental number is the ratio of the circumference to the diameter of a circle (in Euclidean spaces). Like e, π is ubiquitous in mathematics. Famously, π along with four more of the most interesting numbers in mathematics are tied together in the Swiss mathematician Leonhard Euler's remarkable identity: $e^{i\pi} + 1 = 0$.

- 6: the first perfect number.

- $\aleph_1$: this is the first cardinal number greater than the cardinality of the natural numbers (whose cardinality is $\aleph_0$). It is an independent question of ZFC set theory whether $\aleph_1$ is the cardinality of the real numbers (whose cardinality is $2^{\aleph_0}$).

- ω: this is the least infinite ordinal number and, like its cardinal counterpart, $\aleph_0$, it is countable. Unlike its cardinal counterpart, however, there are infinitely many countable infinite ordinals: $\omega, \omega + 1, \omega + 2, ..., \omega \cdot 2, \omega \cdot 2 + 1, \omega \cdot 2 + 2, ..., \omega^2, \omega^3, ..., \omega^{\omega}...$ In ordinal arithmetic, ω is far from the end – it's where things begin to get interesting.

- i: the purely imaginary number $i = \sqrt{-1}$ is fascinating. The Prussian-born German mathematician Leopold Kronecker (1823–91) once said 'God made the integers; all else is the work of man.' But I'm inclined to think that if there were a God and if he or she were in the business of making anything, it would have been i and the rest of the complex numbers at 7.00 a.m. on the first day.

Recommended further reading

Below are a few, mostly mathematical, works where the interested student can pursue some of the material touched upon in this chapter.

Ahlfors, L. V. 1966. *Complex Analysis: An Introduction to the Theory of Analytic Functions of One Complex Variable*, New York: McGraw-Hill. [For more on the Residue Theorem and on complex analysis more generally.]

---

[15] The Fibonacci sequence is, after all, a particular growth series.

Bold, B. 1982. 'The Problem of Squaring the Circle', in *Famous Problems of Geometry and How to Solve Them*, New York: Dover, pp. 39–48. [For more on Lindemann's Theorem and the problem of squaring the circle.]

Colyvan, M. 2006. 'No Expectations', *Mind*, 115(459): 695–702. [A philosophical paper discussing an application of the Riemann Rearrangement Theorem in decision theory.]

Derbyshire, J. 2003. *Prime Obsession: Bernhard Riemann and the Greatest Unsolved Problem in Mathematics*, Washington, DC: Joseph Henry Press. [For more on the Riemann hypothesis and the Prime Number Theorem.]

Devlin, K. 2002. *The Millennium Problems: The Seven Greatest Unsolved Mathematical Puzzles of Our Time*, New York: Basic Books. [For more on the Poincaré conjecture and the Riemann hypothesis.]

Hájek, A. 2010. 'Interpretations of the Probability Calculus', in E. N. Zalta (ed.), *The Stanford Encyclopedia of Philosophy* (Spring 2010 edn), http://plato.stanford.edu/archives/sum2003/entries/probability-interpret/. [An introductory philosophical discussion of the interpretations of probability theory.]

Hallerberg, A. E. 1977. 'Indiana's Squared Circle', *Mathematics Magazine*, 50: 136–40. [For more on the curious history of the Indiana legislature incident involving π.]

Hardy, G. H. 1967. *A Mathematician's Apology*, Cambridge University Press. (First published in 1940.) [A delightful and insightful essay by a leading number theorist.]

Joyce, J. 2008. 'Bayes' Theorem', in E. N. Zalta (ed.), *The Stanford Encyclopedia of Philosophy* (Fall 2008 edn), http://plato.stanford.edu/archives/fall2008/entries/bayes-theorem/. [An introductory philosophical discussion of Bayes's Theorem.]

Kosniowski, C. 1980. *A First Course in Algebraic Topology*, Cambridge University Press. [A good introductory text on algebraic topology; for more on the Borsuk–Ulam Theorem and related topics in topology.]

Millman, R. S. and Parker, G. D. 1977. *Elements of Differential Geometry*, Englewood Cliffs, NJ: Prentice-Hall. [For more on Gauss's Theorema Egregium.]

Nover, H. and Hájek, A. 2004. 'Vexing Expectation', *Mind*, 113(450): 237–49. [A philosophical paper discussing an application of the Riemann Rearrangement Theorem in decision theory.]

Paulos, J. A. 1992. *Beyond Numeracy: An Uncommon Dictionary of Mathematics*, London: Penguin Books. [A good popular book on mathematics with some interesting mathematical odds and ends.]

Rosen, K. H. 2010. *Elementary Number Theory and Its Applications*, 6th edn, Upper Saddle River, NJ: Addison Wesley. [A good introductory text on number theory; for more on the various number theory results discussed in this epilogue.]

Singh, S. 1997. *Fermat's Last Theorem: The Story of a Riddle that Confounded the World's Greatest Minds for 358 Years*. London: Fourth Estate. [A popular book on Fermat's Last Theorem.]

Spivak, M. 2006. *Calculus*, 3rd edn, Cambridge University Press. [A classic introductory calculus text; for more on the Riemann Rearrangement Theorem, the Fundamental Theorem of Calculus, and others.]

Wilson, R. 2002. *Four Colors Suffice*. London: Penguin Books. [For more on the Four-Colour Theorem.]

Bibliography

Ahlfors, L. V. 1966. *Complex Analysis: An Introduction to the Theory of Analytic Functions of One Complex Variable*, New York: McGraw-Hill.

Azzouni, J. 1997. 'Applied Mathematics, Existential Commitment and the Quine–Putnam Indispensability Thesis', *Philosophia Mathematica*, 5(2): 193–227.

2000. 'Applying Mathematics: An Attempt to Design a Philosophical Problem', *The Monist*, 83: 209–27.

2004. *Deflating Existential Consequence: A Case for Nominalism*, Oxford University Press.

Baker, A. 2005. 'Are There Genuine Mathematical Explanations of Physical Phenomena?', *Mind*, 114: 223–38.

2009a. 'Mathematical Explanation in Science', *British Journal for the Philosophy of Science*, 60: 611–33.

2009b. 'Mathematical Accidents and the End of Explanation', in Bueno and Linnebo (eds.), pp. 137–59.

2009c. 'Non-Deductive Methods in Mathematics', in E. N. Zalta (ed.), *The Stanford Encyclopedia of Philosophy* (Fall 2009 edn), http://plato.stanford.edu/archives/fall2009/entries/mathematics-nondeductive/.

2010. 'Mathematical Induction and Explanation', *Analysis*, 70: 681–9.

Balaguer, M. 1996, 'Towards a Nominalization of Quantum Mechanics', *Mind*, 105(418): 209–26.

1998. *Platonism and Anti-Platonism in the Philosophy of Mathematics*, New York: Oxford University Press.

2009. 'Fictionalism, Theft, and the Story of Mathematics', *Philosophia Mathematica*, 17: 131–62.

Batterman, R. W. 2002a. *The Devil in the Details: Asymptotic Reasoning in Explanation, Reduction, and Emergence*, New York: Oxford University Press.

2002b. 'Asymptotics and the Role of Minimal Models', *British Journal for the Philosophy of Science*, 53(1): 21–38.

2010. 'On the Explanatory Role of Mathematics in Empirical Science', *British Journal for the Philosophy of Science*, 61(1): 1–25.

Bays, T. 2009. 'Skolem's Paradox', in E. N. Zalta (ed.), *The Stanford Encyclopedia of Philosophy* (Spring 2009 edn), http://plato.stanford.edu/archives/spr2009/entries/paradox-skolem/.

Beall, J. C. 1999. 'From Full-Blooded Platonism to Really Full-Blooded Platonism', *Philosophia Mathematica*, 7(3): 322–5.

Beall, J. C. and van Fraassen, B. C. 2003. *Possibilities and Paradox*, Oxford University Press.

Benacerraf, P. 1983a. 'What Numbers Could Not Be', in Benacerraf and Putnam (eds.), pp. 272–94.

1983b. 'Mathematical Truth', in Benacerraf and Putnam (eds.), pp. 403–20.

Benacerraf, P. and Putnam, H. (eds.), 1983. *Philosophy of Mathematics: Selected Readings*, 2nd edn, Cambridge University Press.

Bigelow, J. 1988. *The Reality of Numbers: A Physicalist's Philosophy of Mathematics*, Oxford: Clarendon Press.

Bold, B. 1982. 'The Problem of Squaring the Circle', in *Famous Problems of Geometry and How to Solve Them*, New York: Dover, pp. 39–48.

Boolos, G. 1987. 'The Consistency of Frege's Foundations of Arithmetic', in J. Thomson (ed.), *On Being and Saying: Essays for Richard Cartwright*, Cambridge, MA: MIT Press, pp. 3–20, reprinted in Boolos 1998, pp. 183–201.

1998. *Logic, Logic, and Logic*, Cambridge, MA: Harvard University Press.

Borowski, E. J. and Borwein, J. M. 2002. *Collins Dictionary of Mathematics*, 2nd edn, Glasgow: HarperCollins Publishers.

Bostock, D. 2009. *Philosophy of Mathematics: An Introduction*, Oxford: Wiley-Blackwell.

Brouwer, L. E. J. 1983. 'Intuitionism and Formalism', in Benacerraf and Putnam (eds.), pp. 77–89.

Brown, J. R. 2008. *The Philosophy of Mathematics: A Contemporary Introduction to the World of Proofs and Pictures*, 2nd edn, London: Routledge.

Bueno, O. and Colyvan, M. 2011. 'An Inferential Conception of the Application of Mathematics', *Noûs*, 45(2): 345–74.

Bueno, O. and Linnebo, Ø. (eds.) 2009. *New Waves in Philosophy of Mathematics*, Basingstoke, UK: Palgrave Macmillan.

Burgess, J. 1983. 'Why I Am Not a Nominalist', *Notre Dame Journal of Formal Logic*, 24(1): 93–105.

2005. *Fixing Frege*, Princeton University Press.

Burgess, J. P. and Rosen, G. A. 1997. *A Subject with No Object*, Oxford: Clarendon Press.

Buzaglo, M. 2002. *The Logic of Concept Expansion*, Cambridge University Press.

Cajori, F. 1993. *A History of Mathematical Notation*, New York: Dover Reprints. (First published in two volumes by Open Court, London, 1929.)

Colyvan, M. 2001. *The Indispensability of Mathematics*, New York: Oxford University Press.

    2002. 'Mathematics and Aesthetic Considerations in Science', *Mind*, 111: 69–74.

    2005. 'Ontological Independence as the Mark of the Real', *Philosophia Mathematica*, 13(2): 216–25.

    2006. 'No Expectations', *Mind*, 115(459): 695–702.

    2008. 'The Ontological Commitments of Inconsistent Theories', *Philosophical Studies*, 141(1): 115–23.

    2009. 'Applying Inconsistent Mathematics', in Bueno and Linnebo (eds.), pp. 160–72, reprinted in M. Pitici (ed.), *The Best Writing on Mathematics 2010*, Princeton University Press, 2011, pp. 346–57.

    2010. 'There Is No Easy Road to Nominalism', *Mind*, 119(474): 285–306.

Colyvan, M. and Ginzburg, L. R. 2010. 'Analogical Thinking in Ecology: Looking beyond Disciplinary Boundaries', *Quarterly Review of Biology*, 85(2): 171–82.

Conway, J. H. 1976. *On Numbers and Games*, New York: Academic Press.

Courant, R. and Robbins, H. 1978. *What Is Mathematics? An Elementary Approach to Ideas and Methods*, Oxford University Press.

Curry, H. B. 1951. *Outlines of a Formalist Philosophy of Mathematics*, Amsterdam: North Holland.

Davidson, D. 1978. 'What Metaphors Mean', in S. Sacks (ed.), *On Metaphor*, University of Chicago Press, pp. 29–46.

Davies, P. 1992. *The Mind of God*. London: Penguin Books.

Davis, P. J. and Hersch, R. 1981. *The Mathematical Experience*, Boston: Berkhäser.

Dawson, J. W., Jr. 2005. *Logical Dilemmas: The Life and Work of Kurt Gödel*, Natick, MA: A. K. Peters Publishers.

Derbyshire, J. 2003. *Prime Obsession: Bernhard Riemann and the Greatest Unsolved Problem in Mathematics*, Washington, DC: Joseph Henry Press.

De Cruz, H. and De Smedt, J. Forthcoming. 'Mathematical Symbols as Epistemic Actions', *Synthese*.

Devlin, K. 2002. *The Millennium Problems: The Seven Greatest Unsolved Mathematical Puzzles of Our Time*, New York: Basic Books.

Duhem, P. 1954. *The Aim and Structure of Physical Theory*, Princeton University Press. (First published in 1906.)

Dummett, M. 1983. 'The Philosophical Basis of Intuitionism', in Benacerraf and Putnam (eds.), pp. 97–129.

Dyson, F. J. 1964. 'Mathematics in the Physical Sciences', *Scientific American*, 211(3): 128–46.

Eddington, A. 1939. *The Philosophy of Physical Science*, Cambridge University Press.

Eklund, M. 2007. 'Fictionalism', in E. N. Zalta (ed.) *The Stanford Encyclopedia of Philosophy* (Summer 2007 edn), http://plato.stanford.edu/archives/sum2007/entries/fictionalism/.

Enderton, H. B. 1977. *Elements of Set Theory*, New York: Academic Press.

Field, H. 1980. *Science without Numbers*, Oxford: Blackwell.

   1989. *Realism, Mathematics and Modality*, Oxford: Blackwell.

   1993. 'The Conceptual Contingency of Mathematical Objects', *Mind*, 102(406): 285–99.

Franks, C. 2009. *The Autonomy of Mathematical Knowledge: Hilbert's Program Revisited*, Cambridge University Press.

Frege, G. 1967. *The Basic Laws of Arithmetic*, trans. M. Furth, Berkeley: University of California Press.

   1974. *The Foundations of Arithmetic*, trans. J. L. Austin, Oxford: Basil Blackwell.

Friend, M. 2007. *Introducing Philosophy of Mathematics*, Montreal: McGill-Queen's University Press.

George, A. and Velleman, D. J. 2002. *Philosophies of Mathematics*, Malden, MA: Blackwell.

Giaquinto, M. 2002. *The Search for Certainty: A Philosophical Account of the Foundations of Mathematics*, Oxford University Press.

   2007. *Visual Thinking in Mathematics*, Oxford University Press.

Ginzburg, L. R. and Colyvan, M. 2004. *Ecological Orbits: How Planets Move and Populations Grow*, New York: Oxford University Press.

Gödel, K. 1983. 'What is Cantor's Continuum Problem?' (revised and expanded), in Benacerraf and Putnam (eds.), pp. 470–85.

   1992. *On Formally Undecidable Propositions of Principia Mathematica and Related Systems*, New York: Dover.

Gowers, T. (ed.) 2008. *The Princeton Companion to Mathematics*, Princeton University Press.

Gowers, T. and Neilson, M. 2009. 'Massively Collaborative Mathematics', *Nature*, 461 (15 October): 879–81.

Grattan-Guinness, I. (ed.) 2003. *Companion Encyclopedia of the History and Philosophy of the Mathematical Sciences*, Baltimore, MD: Johns Hopkins University Press.

   2007. *The Rainbow of Mathematics: A History of the Mathematical Sciences*, New York: Norton.

2008. 'Solving Wigner's Mystery: The Reasonable (Though Perhaps Limited) Effectiveness of Mathematics in the Natural Sciences', *Mathematical Intelligencer*, 30: 7–17.

Hafner, J. and Mancosu, P. 2005. 'The Varieties of Mathematical Explanation', in P. Mancosu, K. F. Jørgensen, and S. A. Pedersen (eds.), *Visualization, Explanation and Reasoning Styles in Mathematics*, Dordrecht: Springer, pp. 215–50.

Hájek, A. 2010. 'Interpretations of the Probability Calculus', in E. N. Zalta (ed.), *The Stanford Encyclopedia of Philosophy* (Spring 2010 edn), http://plato.stanford.edu/archives/sum2003/entries/probability-interpret/.

Hale, B. and Wright, C. 2001. *The Reason's Proper Study: Essays Towards a Neo-Fregean Philosophy of Mathematics*, Oxford University Press.

Hallerberg, A. E. 1977. 'Indiana's Squared Circle', *Mathematics Magazine*, 50: 136–40.

Hallett, M. 1990. 'Physicalism, Reductionism and Hilbert', in A. D. Irvine (ed.), *Physicalism in Mathematics*, Dordrecht: Kluwer, pp. 183–257.

Hamming, R. W. 1980. 'The Unreasonable Effectiveness of Mathematics', *American Mathematical Monthly*, 87(2): 81–90.

Hardy, G. H. 1967. *A Mathematician's Apology*, Cambridge University Press. (First published in 1940.)

Hart, W. D. 1977. 'Review of Steiner's *Mathematical Knowledge*', *Journal of Philosophy*, 74: 118–29.

(ed.) 1996. *The Philosophy of Mathematics*, Oxford University Press.

Hellman, G. 1989. *Mathematics without Numbers: Towards a Modal-Structural Interpretation*, Oxford: Clarendon Press.

Hersch, R. 1990. 'Inner Vision Outer Truth', in R. E. Mickens (ed.), *Mathematics and Science*, Singapore: World Scientific Press, pp. 64–72.

Herstein, I. N. 1996. *Abstract Algebra*, 3rd edn, New York: Wiley.

Heyting, A. 1971. *Intuitionism: An Introduction*, 3rd rev. edn, Amsterdam: North Holland.

1983. 'The Intuitionist Foundations of Mathematics', in Benacerraf and Putnam (eds.), pp. 52–60.

Hilbert, D. 1983. 'On the Infinite' in Benacerraf and Putnam (eds.), pp. 183–201.

Hofstadter, D. 1979. *Gödel, Escher, Bach: An Eternal Golden Braid*, New York: Basic Books.

Inchausti, P. and Ginzburg, L. R. 2009. 'Maternal Effects Mechanism of Population Cycling: A Formidable Competitor to the Traditional Predator–Prey View', *Philosophical Transactions of the Royal Society B*, 364: 1117–24.

Jacquette, D. (ed.) 2001. *Philosophy of Mathematics: An Anthology*, Malden, MA: Blackwell.

Joyce, J. 2008. 'Bayes' Theorem', in E. N. Zalta (ed.), *The Stanford Encyclopedia of Philosophy* (Fall 2008 edn), http://plato.stanford.edu/archives/fall2008/entries/bayes-theorem/.

Kennedy, J. 2010. 'Kurt Gödel', in E. N. Zalta (ed.), *The Stanford Encyclopedia of Philosophy* (Fall 2010 edn), http://plato.stanford.edu/archives/fall2010/entries/goedel/.

Kitcher, P. 1984. *The Nature of Mathematical Knowledge*, New York: Oxford University Press.

Kline, M. 1972. *Mathematical Thought from Ancient to Modern Times*, New York: Oxford University Press.

Kosniowski, C. 1980. *A First Course in Algebraic Topology*, Cambridge University Press.

Lakatos, I. 1976. *Proofs and Refutations: The Logic of Mathematical Discovery*, Cambridge University Press.

Lange, M. 2009. 'Why Proofs by Mathematical Induction Are Generally Not Explanatory', *Analysis*, 69(2): 203–11.

2010. 'What Are Mathematical Coincidences (and Why Does It Matter)?', *Mind*, 119(474): 307–40.

Leng, M. 2002. 'What's Wrong with Indispensability? (Or the Case for Recreational Mathematics)', *Synthese*, 131: 395–417.

2010. *Mathematics and Reality*, Oxford University Press.

Leng, M., Paseau, A., and Potter, M. 2007. *Mathematical Knowledge*, Oxford University Press.

Levins, R. 1966. 'The Strategy of Model Building in Population Biology', *American Scientist*, 54: 421–31.

Lucas, J. R. 1961. 'Minds, Machines and Gödel', *Philosophy*, 36: 112–27.

Lyon, A. and Colyvan, M. 2008. 'The Explanatory Power of Phase Spaces', *Philosophia Mathematica*, 16(2): 227–43.

MacBride, F. 2003. 'Speaking with Shadows: A Study of Neo-Logicism', *British Journal for the Philosophy of Science*, 54: 103–63.

Maddy, P. 1990. *Realism in Mathematics*, Oxford: Clarendon Press.

1992. 'Indispensability and Practice', *Journal of Philosophy*, 89(6): 275–89.

1995. 'Naturalism and Ontology', *Philosophia Mathematica*, 3(3): 248–70.

1997. *Naturalism in Mathematics*, Oxford: Clarendon Press.

2007. *Second Philosophy: A Naturalistic Method*, Oxford University Press.

Malament, D. 1982. 'Review of Field's *Science without Numbers*', *Journal of Philosophy*, 79: 523–34.

Mancosu, P. (ed.) 2008a. *The Philosophy of Mathematical Practice*, Oxford University Press.

2008b. 'Mathematical Explanation: Why It Matters', in Mancosu (ed.), pp. 134–49.

2008c. 'Explanation in Mathematics', in E. N. Zalta (ed.), *The Stanford Encyclopedia of Philosophy* (Fall 2008 edn), http://plato.stanford.edu/archives/fall 2008/entries/mathematics-explanation/.

Mares, E. D. 2004. *Relevant Logic: A Philosophical Interpretation*, Cambridge University Press.

2009. 'Relevance Logic', E. N. Zalta (ed.), *The Stanford Encyclopedia of Philosophy*, (Spring 2009 edn), http://plato.stanford.edu/archives/spr2009/entries/logic-relevance/.

Maslow, A. H. 1966. *The Psychology of Science: A Reconnaissance*, New York: Harper & Row.

Massey, W. S. 1989. *Algebraic Topology: An Introduction*, New York: Springer-Verlag.

May, R. M. 2004. 'Uses and Abuses of Mathematics in Biology', *Science*, 303(6 February): 790–3.

Melia, J. 2000. 'Weaseling Away the Indispensability Argument', *Mind*, 109: 453–79.

2002. 'Reply to Colyvan', *Mind*, 111: 75–9.

Meyer, R. K. 1976. 'Relevant Arithmetic', *Bulletin of the Section of Logic of the Polish Academy of Sciences*, 5: 133–7.

Meyer, R. K. and Mortensen, C. 1984. 'Inconsistent Models for Relevant Arithmetic', *Journal of Symbolic Logic*, 49: 917–29.

Millman, R. S. and Parker, G. D. 1977. *Elements of Differential Geometry*, Englewood Cliffs, NJ: Prentice-Hall.

Mortensen, C. 1995. *Inconsistent Mathematics*, Dordrecht: Kluwer.

1997. 'Peeking at the Impossible', *Notre Dame Journal of Formal Logic*, 38(4): 527–34.

2004. 'Inconsistent Mathematics', in E. N. Zalta (ed.), *The Stanford Encyclopedia of Philosophy* (Fall 2004 edn), http://plato.stanford.edu/archives/fall 2004/entries/mathematics-inconsistent/.

Munstersbjorn, M. M. 1999. 'Naturalism, Notation, and the Metaphysics of Mathematics', *Philosophia Mathematica*, 7(2): 178–99.

Neumann, J. von 1983. 'The Formalist Foundations of Mathematics', in Benacerraf and Putnam (eds.), pp. 61–5.

Nover, H. and Hájek, A. 2004. 'Vexing Expectation', *Mind*, 113(450): 237–49.

Parsons, C. 1980. 'Mathematical Intuition', *Proceedings of the Aristotelian Society*, 80: 145–68.

Paulos, J. A. 1992. *Beyond Numeracy: An Uncommon Dictionary of Mathematics*, London: Penguin Books.

Penrose, L. S. and Penrose, R. 1958. 'Impossible Objects, a Special Kind of Illusion', *British Journal of Psychology*, 49: 31–3.

Penrose, R. 1989. *The Emperor's New Mind: Concerning Computers, Minds, and the Laws of Physics*, Oxford University Press.

Pincock, C. 2004. 'A New Perspective on the Problem of Applying Mathematics', *Philosophia Mathematica*, 12: 135–61.

  2007. 'A Role for Mathematics in the Physical Sciences', *Noûs*, 41: 253–75.

Priest, G. 1997. 'Inconsistent Models of Arithmetic Part I: Finite Models', *Journal of Philosophical Logic*, 26(2): 223–35.

  2000. 'Inconsistent Models of Arithmetic Part II: The General Case', *Journal of Symbolic Logic*, 65: 1519–29.

  2005. *Towards Non-Being: The Logic and Metaphysics of Intentionality*, Oxford: Clarendon Press.

  2008. *An Introduction to Non-Classical Logic: From If to Is*, 2nd edn, Cambridge University Press.

Priest, G. and Tanaka, K. 2004. 'Paraconsistent Logic', in E. N. Zalta (ed.), *The Stanford Encyclopedia of Philosophy* (Winter 2004 edn), http://plato.stanford.edu/archives/win2004/entries/logic-paraconsistent/.

Putnam, H. 1971. *Philosophy of Logic*, New York: Harper, reprinted in *Mathematics, Matter and Method: Philosophical Papers, vol. I*, 2nd edn, Cambridge University Press, 1979, pp. 323–57.

  1979. 'What is Mathematical Truth?', in *Mathematic, Matter and Method: Philosophical Papers, vol. I*, 2nd edn, Cambridge University Press, pp. 60–78.

  1980. 'Models and Reality', *Journal of Symbolic Logic*, 45(3): 464–82.

Quine, W. V. 1953. 'Two Dogmas of Empiricism', in *From a Logical Point of View*, Cambridge, MA: Harvard University Press, pp. 20–46.

  1960. *Word and Object*, Cambridge, MA: MIT Press.

  1976. 'Carnap and Logical Truth', in *The Ways of Paradox and Other Essays*, rev. edn, Cambridge, MA: Harvard University Press, pp. 107–32 (and in Benacerraf and Putnam (eds.), pp. 355–76).

  1981a. 'Five Milestones of Empiricism', in *Theories and Things*, Cambridge, MA: Harvard University Press, pp. 67–72.

  1981b. 'Success and Limits of Mathematization', in *Theories and Things*, Cambridge, MA: Harvard University Press, pp. 148–55.

  1986. 'Reply to Charles Parsons', in L. Hahn and P. Schilpp (eds.), *The Philosophy of W. V. Quine*, La Salle, IL: Open Court, pp. 396–403.

Resnik, M. D. 1985. 'How Nominalist Is Hartry Field's Nominalism?', *Philosophical Studies*, 47: 163–81.

    1995. 'Scientific vs. Mathematical Realism: The Indispensability Argument', *Philosophia Mathematica*, 3(2): 166–74.

    1997. *Mathematics as a Science of Patterns*, Oxford: Clarendon Press.

Resnik, M. D. and Kushner, D. 1987. 'Explanation, Independence, and Realism in Mathematics', *British Journal for the Philosophy of Science*, 38: 141–58.

Robinson, A. 1966. *Non-standard Analysis*, Amsterdam: North Holland.

Rosen, K. H. 2010. *Elementary Number Theory and Its Applications*, 6th edn, Upper Saddle River, NJ: Addison Wesley.

Shapiro, S. 1997. *Philosophy of Mathematics: Structure and Ontology*, New York: Oxford University Press.

    2000. *Thinking about Mathematics: The Philosophy of Mathematics*, Oxford University Press.

    2005. *Oxford Handbook of Philosophy of Mathematics and Logic*, New York: Oxford University Press.

Sieg, W. 1999. 'Hilbert's Programs: 1917–1922', *Bulletin of Symbolic Logic*, 5(1): 1–44.

Singh, S. 1997. *Fermat's Last Theorem: The Story of a Riddle that Confounded the World's Greatest Minds for 358 Years*, London: Fourth Estate.

Skolem, T. 1922. 'Some Remarks on Axiomatized Set Theory', in J. van Heijenoort (ed.), *From Frege to Gödel: A Source Book in Mathematical Logic 1879–1931*, Cambridge, MA: Harvard University Press, 1967, pp. 290–301.

Smart, J. J. C. 1990. 'Explanation – Opening Address', in D. Knowles (ed.), *Explanation and Its Limits*, Cambridge University Press, pp. 1–19.

Smith, P. 2007. *An Introduction to Gödel's Theorems*, Cambridge University Press.

Sober, E. 1993. 'Mathematics and Indispensability', *Philosophical Review*, 102(1): 35–57.

Sorensen, R. A. 1988. *Blindspots*, Oxford: Clarendon Press.

Spivak, M. 2006. *Calculus*, 3rd edn, Cambridge University Press.

Steiner, M. 1978a. 'Mathematical Explanation', *Philosophical Studies*, 34: 135–51.

    1978b. 'Mathematics, Explanation, and Scientific Knowledge', *Noûs*, 12: 17–28.

    1989. 'The Application of Mathematics to Natural Science', *Journal of Philosophy*, 86: 449–80.

    1995. 'The Applicabilities of Mathematics', *Philosophia Mathematica*, 3: 129–56.

    1998. *The Applicability of Mathematics as a Philosophical Problem*, Cambridge, MA: Harvard University Press.

Thomasson, A. L. 1999. *Fiction and Metaphysics*, Cambridge University Press.

Tymoczko, T. 1998 *New Directions in the Philosophy of Mathematics*, rev. and expanded paperback edn, Princeton University Press.

Urquhart, A. 1990, 'The Logic of Physical Theory', in A. D. Irvine (ed.), *Physicalism in Mathematics*, Dordrecht: Kluwer, pp. 145–54.

Weber, Z. Forthcoming. 'Figures, Formulae, and Functors', in S. Shin and A. Moktefi (eds.), *Visual Reasoning with Diagrams*, Springer.

Weinberg, S. 1986. 'Lecture on the Applicability of Mathematics', *Notices of the American Mathematical Society*, 33: 725–8.

    1993. *Dreams of a Final Theory*, London: Vintage.

Weir, A. 2010. *Truth through Proof: A Formalist Foundation for Mathematics*, Oxford University Press.

    2011. 'Formalism in the Philosophy of Mathematics', in E. N. Zalta (ed.), *The Stanford Encyclopedia of Philosophy* (Spring 2011 edn), http://plato.stanford.edu/archives/spr2011/entries/formalism-mathematics/.

Whitehead, A. N. and Russell, B. 1910, 1912, 1913. *Principia Mathematica*, 3 vols., Cambridge University Press; 2nd edn, 1925 (vol. 1), 1927 (vols. 2, 3).

Wigner, E. P. 1960. 'The Unreasonable Effectiveness of Mathematics in the Natural Sciences', *Communications on Pure and Applied Mathematics*, 13: 1–14.

Wilson, M. 2000. 'The Unreasonable Uncooperativeness of Mathematics in the Natural Sciences', *The Monist*, 83: 296–315.

Wilson, R. 2002. *Four Colours Suffice: How the Map Problem was Solved*, London: Penguin Books.

Wright, C. 1983. *Frege's Conception of Numbers as Objects*, Aberdeen University Press.

Yablo, S. 1998. 'Does Ontology Rest on a Mistake?', *Proceedings of the Aristotelian Society, Supplementary Volumes*, 72: 229–61.

    2005. 'The Myth of the Seven', in M. Kalderon (ed.), *Fictionalism in Metaphysics*, Oxford: Clarendon Press, pp. 88–115.

    2009. 'Must Existence-Questions Have Answers?', in D. J. Chalmers, D. Manley, and R. Wasserman (eds.), *Metametaphysics: New Essays on the Foundations of Ontology*, Oxford University Press, pp. 507–26.

Zach, R. 2009. 'Hilbert's Program', in E. N. Zalta (ed.), *The Stanford Encyclopedia of Philosophy* (Spring 2009 edn), http://plato.stanford.edu/archives/spr2009/entries/hilbert-program/.

Zalta, E. N. 1983. *Abstract Objects: An Introduction to Axiomatic Metaphysics*, Dordrecht: Reidel.

    1999. 'Natural Numbers and Natural Cardinals as Abstract Objects: A Partial Reconstruction of Frege's Grundgesetze in Object Theory', *Journal of Philosophical Logic*, 28(6): 619–60.

2000. 'Neologicism? An Ontological Reduction of Mathematics to Meta-physics', *Erkenntnis*, 53(1–2): 219–65.

2010. 'Frege's Logic, Theorem, and Foundations for Arithmetic', in E. N. Zalta (ed.), *The Stanford Encyclopedia of Philosophy* (Fall 2010 edn), http://plato.stanford.edu/archives/fall2010/entries/frege-logic/.

Index